Tropospheric Ozone and its Impacts on Crop Plants

Supriya Tiwari • Madhoolika Agrawal

Tropospheric Ozone and its Impacts on Crop Plants

A Threat to Future Global Food Security

 Springer

Supriya Tiwari
Department of Botany, Institute of Science
Banaras Hindu University
Varanasi, UP, India

Madhoolika Agrawal
Department of Botany, Institute of Science
Banaras Hindu University
Varanasi, UP, India

ISBN 978-3-030-10125-1   ISBN 978-3-319-71873-6   (eBook)
https://doi.org/10.1007/978-3-319-71873-6

Printed on acid-free paper

This Springer imprint is published by Springer Nature
The registered company is Springer International Publishing AG
The registered company address is: Gewerbestrasse 11, 6330 Cham, Switzerland

Preface

The presence of ozone in low quantities in the troposphere has been well established since the pre-industrial times. However, sudden unpredicted increase in background concentrations of ozone in the last few decades has alarmed environmentalists worldwide. As a secondary pollutant which is formed by a series of photochemical reactions, ozone concentration in the troposphere depends upon a number of factors, the important ones being meteorological variables and rate of emission of ozone precursors like nitrogen oxides and volatile organic compounds. In the last few years, although developed countries like the USA have managed to bring down their rate of emission of important ozone precursors, background ozone concentration has still registered an increase. The unplanned urbanization and industrialization practised in developing countries, especially those in south and east Asia, have made these regions important hotspots for ozone production in the near future. The recently proposed concept of intercontinental transportation has a significant influence on the tropospheric ozone budget of a region irrespective of the extent of air quality legislations followed in that region. Ozone, therefore, has assumed a global significance rather than remaining confined as a regional problem.

The present scenario of global climate change has predicted significant variations in the different meteorological variables that play major roles in in situ ozone production. As such, climate change has a strong influence on the global tropospheric budget. In addition, climate change also modifies the stratospheric-tropospheric intrusion of ozone which is also an important component in determining tropospheric ozone budget. Apart from ozone formation, the halogen chemistry involved in ozone destruction is also important in establishing the global tropospheric ozone concentration. This book not only discusses the different aspects that control ozone concentration in the troposphere, but also emphasizes the role of climate change in regulating tropospheric ozone budget directly or indirectly.

The phytotoxic effects of tropospheric ozone can be attributed to its oxidizing nature. Ozone has the potential to generate extra reactive oxygen species (ROS) like superoxide dismutase, hydroxyl ions, singlet oxygen, and hydrogen peroxide which disturb the normal metabolism of the cell/plant. Plants have a built-in constitutive enzymatic antioxidant system and a non-enzymatic antioxidant system which are stimulated in response to ROS generation. Under stress conditions, the ROS generation exceeds the scavenging capacity of the cellular defence machinery which disrupts the various biochemical and physiological processes of the cell. The main targets of ozone-induced ROS are the cellular and subcellular membranes whose permeability is affected due to the peroxidation of their lipid components.

Plants exposed to ozone show different symptoms which are indicative of ozone-induced stress and ultimately lead to reduced carbon fixation. In addition to this, the biomass allocation pattern of the ozone-stressed plants is also disturbed, thus affecting its yield components. Several ozone impact assessment programmes and individual experiments conducted worldwide have proved that ozone induces significant yield reductions in important staple food crops. Several modelling studies further suggest that in view of the unchecked continuous increase in ozone concentration, the yield reductions are further bound to increase which may prove to be a serious threat to global food security in the near future.

Since ozone formation in the troposphere requires a diversified setup, it is difficult to check the increasing concentration of ozone around the globe. Therefore, the need of the hour is to develop certain strategies that allow the plants to alleviate or minimize the ozone-induced stress. Carbon dioxide fertilization and soil nutrient amendments are the important techniques discussed in this book that show promising results as far as the yield response of plants under ozone stress is concerned. However, more experimentations and thorough studies related to the mechanism of alleviation of ozone stress should be pursued before actually implementing their use as tools for minimizing ozone injury.

This book provides a valuable overview of issues related to increasing concentration of tropospheric ozone, focussing on the response of plants to ozone stress. Ozone itself is a major environmental problem, and the present climate change scenario has further increased the frequency of occurrence of favourable events promoting ozone formation leading to increased tropospheric ozone budget. This book will be quite useful for the students of various branches of biological and environmental science.

Varanasi, UP, India   Supriya Tiwari
                                                                          Madhoolika Agrawal

Contents

About the Author

Supriya Tiwari is an Assistant Professor in the Department of Botany at Banaras Hindu University, India. Her research focuses on the formation of ozone and its effects on plant productivity in India. She has also worked on the mechanisms of plants' response under ozone stress which lead to yield reductions. Her evaluations of ozone crop injury and ozone-induced yield reductions have made significant contributions in the planning of sustainable agriculture strategies towards maintaining ozone-induced yield losses.

Madhoolika Agrawal is a Professor in the Department of Botany at Banaras Hindu University, India. Her areas of specialization include air pollution impact assessment with emphasis on surface ozone monitoring and evaluation of ozone-induced crop yield losses. Her research findings include the quantification of ozone impacts on global food security. She has been focusing on screening ozone-resistant cultivars, along with the development of a few strategies like the use of nutrient amendments and carbon dioxide fertilization for minimizing ozone injury in plants.

Chapter 1
Ozone Concentrations in Troposphere: Historical and Current Perspectives

Abstract The concentration of ground level ozone (O_3) has registered an unpredictably high increase in the last few decades. It has been observed that the background O_3 concentration has doubled since the nineteenth century with more prominent effects in Northern hemisphere. Although the formation of O_3 largely depends upon the regional emission of O_3 precursors, the increasing O_3 concentrations have acquired a global significance. With the implementation of air quality legislations, the anthropogenic emission of O_3 precursors has declined in North America and Europe but the problem still persists in Asia. The long lifetime of O_3 superimposed by its intercontinental transport from Asia play an important role increasing global background O_3 concentration in North America and Europe. Modeling studies have recognized South and East Asia as the major hotspots where O_3 concentrations are expected to show maximum increase in near future. The present chapter throws light on the historical background of O_3 monitoring and discusses the present scenario of ground level O_3 with emphasis on the recent trends in different continental zones along with the seasonal and diurnal variations.

Keywords Background ozone · Ozone precursors · Monitoring · Hotspots

Contents

© Springer International Publishing AG 2018
S. Tiwari, M. Agrawal, *Tropospheric Ozone and its Impacts on Crop Plants*,
https://doi.org/10.1007/978-3-319-71873-6_1

1   Introduction

The history of tropospheric O_3 can be traced back to the times when oxygen first appeared on the Earth. Ozone was formed as a byproduct of oxygen production in the presence of ultraviolet (UV) radiations reaching the Earth's surface. In the atmosphere, O_3 exists in two different atmospheric zones (Fig. 1.1). The upper zone, the stratosphere extends between 10 and 50 km and has a significantly high concentration of O_3, playing a beneficial role in screening the lower layers of atmosphere from sun's harmful UV radiations. The lower atmospheric zone, the troposphere extends from the Earth's surface up to the stratosphere and the role of O_3 reverses as it behaves as an extremely damaging molecule due to its highly oxidizing nature (Baier et al. 2005; Sharma et al. 2012). This contrasting behavior of O_3 leads to its categorization as "good ozone", which is present in the stratosphere and "bad ozone", present in the troposphere. The thickness of O_3 column in the stratosphere was on continuous decline, however, in the last few decades, due to serious and extensive efforts, stratospheric O_3 depletion has reduced substantially (UNEP 2014). Tropospheric O_3 on the other hand shows a continuous increasing trend (Tiwari et al. 2008; IPCC 2013; Cooper et al. 2014; Lin et al. 2014) and has become an issue of serious concern world over due to its damaging impacts on agricultural crops and vegetation (Embersen et al. 2009; Fowler et al. 2009; Teixeira et al. 2011).

Tropospheric O_3 is also well recognized as an important short lived climate pollutant (Shindell et al. 2012) as well as an important greenhouse gas (Kumar and Imam 2013). O_3 is peculiar and different from other gaseous pollutants as it does not have any specific emission sources (secondary pollutant), but is formed in the atmosphere by a few photochemical reactions involving certain chemical moieties called O_3 precursors, such as carbon monoxide (CO), volatile organic compounds (VOCs) and nitrogen oxides (NOx) (The Royal Society 2008; Monks et al. 2015). Tropospheric ozone budget at a particular location depends on the number of factors such as proximity to precursor sources, geographical location and prevailing meteorological conditions (Khiem et al. 2010; Monks et al. 2009). These precursors arise from both natural as well as anthropogenic sources (Crutzen 1973; Chameides and Walker 1973). Initially, it was beleived that the primary source of ground level O_3

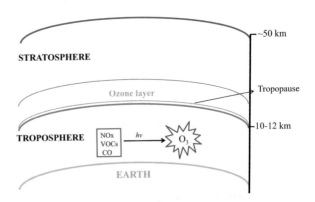

Fig. 1.1 Atmospheric stratification into troposphere and stratosphere

was only the downward transport of O_3 from the stratosphere to the troposphere (Junge 1962). However, extensive researches conducted during 1970s and 1980s linked the origin of O_3 to a few photochemical reactions occurring in the atmosphere near ground level (Crutzen 1974; Logan et al. 1981; Liu et al. 1987). Photochemical mechanisms for O_3 formation were first identified in California during 1950s (Haagen-Smit 1952). In 1970s, Crutzen (1973) and Chameides and Walker (1973) explained the production of tropospheric O_3 by the photochemical oxidation of CO and hydrocarbons. Since then, the photochemical processes were established to be the major sources of O_3 in the troposphere, rather than the earlier assumption which gave more emphasis to downward transport of O_3 from the stratosphere.

Present day Chemical Transport and Chemistry Climate Global models show considerable variations in the percentage of O_3 originating from downward stratospheric transport or from in situ photochemistry, but agree that photochemical production is the main source of ground level O_3 (Wu et al. 2007). O_3 production by photochemical reactions exceed its stratospheric transport by a factor ranging from 7 to 15 (Young et al. 2013). Model assessments suggest that chemical production contributes about 5000 Tg yr^{-1} to Global tropospheric O_3 (Stevenson et al. 2006), in contrast to the Stratospheric-Tropospheric Exchange (STE), which contributes merely 550 ± 140 Tg yr^{-1} (Olsen et al. 2004; McLinden et al. 2000). Young et al. (2013) suggested that the total O_3 burden increased by ~30% since mid twentieth century, 30% of which is attributed to anthropogenic emissions.

With the establishment of the significance of photochemical reactions in O_3 formation, it can be suggested that the concentrations of O_3 largely depend upon the concentrations of its precursors and the climatic conditions (Lee et al. 2014). Increased emission of precursors and favourable climatic conditions lead to high concentrations of O_3 in different regions of the globe (IPCC 2013; Monks et al. 2009). Thus, due to the variable sources of O_3 precursors, role of natural and physical processes and complex chemistry of O_3 formation, checking the increasing concentrations of O_3 is not an easy and straightforward task. The only practically feasible strategy is to control the anthropogenic activities that emit precursors for O_3 formation (The Royal Society 2008; Lin et al. 2008; Monks et al. 2009). At global scale, O_3 pollution is highest in Central Europe, Eastern China, Eastern USA and South East Asia.

2 Pre-historic Ozone Concentrations

O_3 has always existed at background concentrations which vary geographically and throughout the year. The first measurement of O_3 can be traced to early nineteenth century, when Christian F. Schönbein, Professor of Chemistry at Basal, Switzerland developed a semi-quantitative method of measuring O_3 concentrations in the atmosphere using strips coated with a mixture of starch and KI (Potassium Iodide). The colour developed on this strip after 12 h exposure to air was compared to a 10-grade chromatic scale which gave a near accurate estimate of maximum O_3 abundance during 12 h period. However, the results obtained through this technique largely depended upon humidity, air flow, other oxidants present in the atmosphere and accidental exposure to sunlight. Despite these limitations, motivated by this methodology, around 300 recording stations were established in different parts of Europe and USA, to measure ambient O_3 concentrations in the beginning of 1850s. However, as the continuity of measurements was not maintained, hence long term data are limited. The recorded data indicate about the background concentrations of O_3 in the absence of any significant anthropogenic emissions. However, Schönbien method could only indicate the relative quantity of O_3.

The first quantitative O_3 measurements were taken at Paris Municipal Observatory in Park Montsouris, from 1876 to 1907. Volz and Kley (1988) reanalyzed the Montsouris data and concluded that average O_3 mixing ratios from 1876 to 1910 were 11 ppbv with an uncertainty of ±2 ppbv due to SO_2 corrections that inferred with the measurement techniques. During 1980s and 1990s, several researchers attempted to convert the quantity of O_3 reported by Schönbein in late 19th and early twentieth century, to O_3 mixing ratios using regression between quantitative Montsouris measurements and coincident Schönbien's measurements as developed by Bojkov (1986). Linvill et al. (1980) also established a relationship between O_3 measured through modern UV absorption technique and reconstructed Schönbien papers. However, the O_3 concentration recorded through both these approaches showed significant variations. A major limitation of these approaches is the confounding non-linear sensitivity of Schönbien measurements to humidity. Linvill et al. (1980) reconstructed the Schönbien data and observed that the value of daytime O_3 at Michigan for all months during 1876–1880 was 35 ppbv. However, Bojkov's (1986) approach found O_3 values which were different from Linvill et al. (1980). Marenco et al. (1994) showed Bojkov regression equation to be errorful, which led to higher O_3 estimates. O_3 mixing ratios were estimated through revised Bojkov's method (1986) along with a consideration of Linvill et al. (1980) for late nineteenth century and early twentieth century O_3 measurements conducted at different sites, which included Moncalieri in Northern Italy (1863–1893) (Anfossi et al. 1991), Montevideo, Uruguay (1883–1885) and Cordoba, Argentina (1886–1892) (Sandroni et al. 1992), Athens, Greece (1901–1940) (Cartalis and Varatosis 1994) and Zagreb, Croatia (1889–1900) (Lisac et al. 2010) and Malta in Mediterranean Sea (Nolle et al. 2005). Besides, O_3 estimates were also analyzed at five southern hemisphere and seven northern hemisphere sites at different time spans ranging from 1872 to 1928 (Pavelin et al. 1999). A close analysis of the above studies led to the following conclusions:

(i) In nineteenth century, seasonal O_3 most often peaked in spring, followed by winter (Cooper et al. 2014).

(ii) A comparison of O_3 mixing ratios in nineteenth century with twentieth century indicate that O_3 increased by a factor of about 2 during twentieth century (Cooper et al. 2014; Meehl et al. 2007).

In 1930s, a few short term measurements were recorded at several high altitude sites in Switzerland, Germany and France. The results recorded when compared with the Schönbien measurements done in 1870 showed an exponential increase in

O_3 by a factor of 2 between early 1900s and 1950s (Marenco et al. 1994). Staehelin et al. (1994) used additional data sets from 1930 to 1950 in Central Europe and reached to a similar conclusion and showed an increment in O_3 concentration by a factor of 2.2 in Alpine valley from 1950s to 1989–1991. Parrish et al. (2012) further reported that quantitative measurements from 1870s to 1950s indicate that surface O_3 in Europe increased by a factor of 2. The magnitude of the observable O_3 increments corresponds with the global increase in the fossil fuel combustion as reported in the IPCC Fifth Assessment Report (IPCC 2013). Rising O_3 concentrations can also be explained through the rising NOx emissions in Europe that increased by a factor of 4.5 between 1955 and 1985 (Staehelin et al. 1994). The results of the World's largest and continuous O_3 data recorded from Arkona- Zingst site on Northern German coast carried out from 1956 to 1990, are in agreement with the trends compiled by Marenco et al. (1994) and Staehelin et al. (1994). All these observations suggest that the yearly average value of O_3 ranged between 15 and 20 ppbv during 1990s and 1960s, which doubled by the end of twentieth century (Parrish et al. 2012).

3 Present Scenario of Ozone Concentrations

Studies have shown that O_3 concentration has more burdened across the tropics and mid latitudes of northern hemisphere than in southern hemisphere (Ziemke et al. 2011). It has been observed that the background concentrations of O_3 in mid latitudes of northern hemisphere during nineteenth century has doubled to about 30–35 ppb and have since then increased by another 5 ppb to reach 35–40 ppb (The Royal Society 2008). Modeling studies have suggested that global O_3 concentrations will increase during the early part of the twenty-first century, as a result of increasing precursor emissions especially in Northern mid latitudes, with western North America being particularly sensitive to rising Asian emissions (Cooper et al. 2010). These models take into account, the O_3 precursor emissions, transport/removal of O_3 molecules, atmospheric chemistry, etc. However, it is not possible to simulate O_3 at global, regional and urban levels through a single model. Therefore, different models have been developed for O_3 studies in a wide range of spatial scales. Global models cover the whole troposphere and lower stratosphere as well, with a horizontal resolution of approximately 200 km, for example MOZRAT-2 (Horowitz et al. 2003), ACCENT (18 atmospheric models from US, Europe and Japan) (Ellingsen et al. 2008). Regional models cover spatial scales of a single continent eg. Regional Chemical Transport Model (WRF-Chem) (Jena et al. 2015). National scale models used for smaller region are more detailed eg LUR Model, used to estimate O_3 at 90 locations in Netherland (Kerckhoffs et al. 2015). At even finer scales, urban models at <1 km scales and street canyon models (~1 m) have been developed. ARIMA Model used in predicting ground level O_3 at urban and rural sampling points in South Eastern Spain, is an example of urban model (Duenas et al. 2005).

Since the photochemical processes are established to be the main phenomenon responsible for the increasing O_3 in the troposphere, the concentrations of O_3 precursors are important in determining the tropospheric O_3 budget (Monks et al. 2009). While O_3 has a relatively short atmospheric lifetime, in polluted urban centres which have high concentration of O_3 precursors, lifetime of free tropospheric O_3 extends to several weeks (Young et al. 2013). These observations extend the role of O_3 precursors which is now not limited to O_3 formation but also play an important role in intercontinental O_3 transport. The extended lifetime facilitates the transportation of O_3 over intercontinental scale, thus interfering with the air quality at hemispheric level (HTAP 2010; Simpson et al. 2014). In addition to this, O_3 precursors also control the tropospheric distribution of hydroxyl radicals (Naik et al. 2013; Lin et al. 2014). Hydroxyl radicals control the oxidizing capacity of the troposphere, which in turn regulates the lifetime and build up of methane (Pusede et al. 2015; Derwent et al. 2001).

Ozone, as such is no longer considered as a regional or local problem, but has assumed a much larger, intense and global significance. Changes in regional O_3 levels not only depend upon rate of emission of O_3 precursors and existing meteorological conditions, but also upon the transportation of O_3 from one region to another along with the prevailing winds. Over the past few decades, with the implementation of air quality legislations, anthropogenic emissions of O_3 precursors have declined in North America and Europe, while they are continuously increasing in Asia (Tai et al. 2013; Granier et al. 2011). It has been observed that O_3 can be transported from East Asia to Western North America through the prevailing winds (Doherty 2015). A few studies have also suggested a correlation between O_3 entering the US west coast and the local pollution levels in California (Parrish et al. 2010; Huang et al. 2010).

Being a hemispheric pollutant, reduction in O_3 concentrations at local or National level does not necessarily bring about significant changes in its background concentrations. This is evident from the fact that although in the past few decades, the emissions of most of the O_3 precursors have decreased over Western North America, the concentrations of tropospheric O_3 levels have increased in the spring time (Cooper et al. 2012). An analysis of long term (1990–2010) rural O_3 trends at selected sites of western and eastern USA showed that during spring, 43% of the sites of eastern USA recorded a statistically significant reduction, while 50% of the sites of western USA showed increase in O_3 concentrations (Cooper et al. 2012). O_3 transportation clearly explains the discrepancies in the concentrations of O_3 across the North America. Concentration of ground level O_3 is significantly affected by intercontinental transport through trans-pacific and trans-atlantic pathways of European and Asian O_3 precursor emissions (Doherty 2015). Jaffe et al. (1999) found that the observed ground level O_3 concentrations over northwest US were greatly affected by episodic pollutant plums from Asia. Using the calculations from MOZART-2.4, Lin et al. (2008) observed that European and Asian anthropogenic emission of precursors contributed 4–5 ppb (summer average daily mean) over northwest US, 3–4 ppb over California and less than 2 ppb over eastern US. MOZART-2.4 model study was used to project changes in ground level O_3 concentration over US from 1999 to 2049 and to 2099, due to changes in global climate and in European and Asian natural and anthropogenic emissions leading to 3–8 ppb more O_3 over eastern US (Lin et al. 2008).

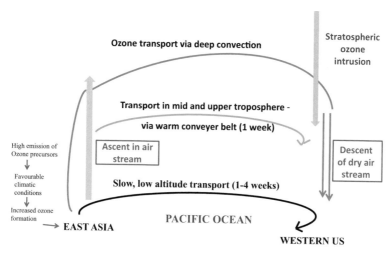

Fig. 1.2 Intercontinental ozone transportation via different modes (Modified from Doherty 2015)

Occurrence of cyclones in the mid latitude regions during winter and spring plays a significant role in O_3 trans-pacific transport (Lin et al. 2012). In winter and spring, the transport of pollutants from East Asia to western North America occurs via mid latitude cyclones that uplift the surface O_3 and its precursors to middle and upper troposphere, followed by descent via dry air entrenched within the cyclone over western North America (Doherty 2015). In summers, weaker cyclones and increased photochemical destruction diminish the impact of East Asian pollution over western North America (Brown-Steiner and Hess 2011). In this condition, deep convection is typically the dominant mechanism of intercontinental O_3 transportation (Doherty 2015). In addition to the anthropogenic activities, downward transport of naturally occurring stratospheric O_3 is also an important source of O_3 in the troposphere. This transport is more prominent through the descending day air streams of mid latitude cyclones and has a maximum impact on free tropospheric O_3 in the springtime (Lin et al. 2015). Slow low altitude transport is another mode prevalent through all the seasons (Fig. 1.2).

O_3 transportation, therefore, plays an important role in determining the global O_3 concentrations. The rising emissions of O_3 precursors in East Asia are not only responsible for increase in the ground level O_3 concentrations in Asia, but also over Western North America during spring season (Parrish et al. 2012; Lin et al. 2012). Cooper et al. (2012) have reported that anthropogenic emissions of O_3 precursors declined by 49, 45 and 44% for NOx, CO and VOC, respectively during 1990–2010. However, O_3 concentrations over Western North America have increased significantly at the rate 0.41 ± 027 ppb yr^{-1} from 1995 to 2011 (Cooper et al. 2012). Studies have shown that the Asian pollution on reaching the Eastern North Pacific Ocean splits in Northern and Southern branches, with O_3 production being more prominent in Southern branch (Zhang et al. 2008; Fischer et al. 2010). Using satellite data and GFDL AM3 model (Donner et al. 2011) simulation studies, Lin et al.

(2012) confirmed the above statement and proposed that O_3 production further extends in western US during spring time and contributes to Asian enhancement of surface O_3 in US. Satellite based observations have shown that pollution plumes not only travel from East Asia to western North America, but can also travel from North America to Europe and can even encircle the globe (Monks et al. 2009).

Studies have shown that intercontinental O_3 transportation has significant effects on surface O_3 concentrations in Asia which is influenced by emissions from Europe. A study of O_3 profile has shown that in Western India, Ahmedabad is particularly affected by emissions and transport from Southern Europe and North Africa (Srivastava et al. 2011; Lal et al. 2013). Intercontinental transport is studied through chemical transport models or chemistry climate models. However, even the improved versions of these models show a monthly mean difference of 10 ppb in the quantity of background O_3 in western North America (Fiore et al. 2014). Future climate change predictions suggest larger impacts on USA from East Asian emissions (Glotfelty et al. 2014).

Verstraeten et al. (2015) compared the annual changes in tropospheric O_3 column over China and United States between 2005 and 2010 using the combination of satellite data and a Chemical Transport model data generated through Global Chemistry Climate Model. Over China, and reported an increase of 21% in emission of O_3 precursor species, NOx was reported with a subsequent increase of 1.08% in the free troposphere O_3 column between altitudes of 3–9 km. In contrast, no significant change in free tropospheric O_3 was reported over US, although NOx emission declined by 21% during 2005–2010 (Verstraeten et al. 2015). This study further suggests that changes in NOx emissions over China were responsible for increase in ground level O_3 concentrations over Western US. The findings of Verstraeten et al. (2015) further strengthened the suggestions of Dentener et al. (2011) that implementation of National Control policies regarding O_3 air quality should be improved with International agreements, especially in downwind areas whose O_3 precursor concentrations may affect the O_3 concentrations in nearby upwind areas.

An assessment of O_3 concentrations was done on global scale (70° N- 70° S) with the help of Ozone Monitoring Instrument (OMI) onboard NASA's polar orbiting Aura satellite, monitoring the tropospheric and stratospheric O_3 since 2004. These observations produced global maps of Tropospheric Column Ozone (TCO) (Ziemke et al. 2006). OMI related TCO suggests that northern hemisphere experiences its peak O_3 values during summers at mid latitude, while the southern hemisphere peaks in spring especially in tropical and subtropical regions (Ziemke et al. 2011) (Fig. 1.3). On an annual basis, the northern hemisphere's average TCO exceeds the southern hemisphere's averages by 4, 12 and 18% at low (0–25°), mid (25–50°) and high (50–60°) latitudes, respectively (Ziemke et al. 2011). Comparing the two hemispheres, it was observed that 52% of the annual average tropospheric O_3 is recorded from the northern hemisphere, while 48% resides in southern hemisphere (Ziemke et al. 2011). The lower surface O_3 values in southern hemisphere mid latitudes were 24–25 ppbv from Ushuaia, Argentina, 24–26 ppbv from Cape point, South Africa, 24–26 ppbv from Cape Grim, Tasmania and 27–28 ppbv from Arrival Heights, Antarctica. The lowest O_3 concentrations have been recorded in Marine Boundry Layer (MBL) of tropical southern hemisphere (Samoa, 13–14 ppbv) (Cooper et al. 2014). The northern hemisphere, however,

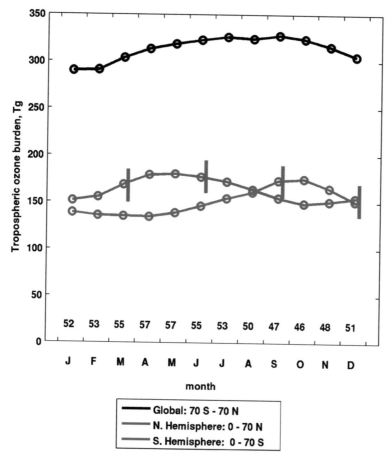

Fig. 1.3 Global monthly variations in tropospheric ozone (70° S–70° N) in Northern and Southern Hemispheres (Source: Cooper et al. 2014)

had high O_3 values ranging from 38 to 46 ppbv in tropics (Mauna Loa), 37–43 ppbv in southern Germany, 40–44 ppbv in northern California and 45–48 ppbv in Greenland. Lowest O_3 concentration in northern hemisphere, comparable to southern hemisphere is found at the coastal site of Barrow, Alaska (27–28 ppbv) (Oltmans et al. 2012).

The greatest O_3 mixing ratio in northern hemisphere is found in Western Japan, immediately downwind of continent East Asia, where O_3 precursor emissions have increased significantly since 1980 (Granier et al. 2011; Lee et al. 2014). Through recent studies based on satellite observations made by TOMS (Total O_3 Mapping Spectrometer), Cooper and Ziemke (2013) observed that between 2005 and 2014, O_3 burden across the globe (60°S- 60°N) increased by 0.71% yr^{-1}. Ebojie et al. (2016) analyzed the global variations of tropospheric O_3 columns (TOCs) derived from SCIAMACHY limb-nadir matching (LNM) observations. During the period 2003–2011, O_3 showed an increase of 0.1% yr^{-1} in southern hemisphere (50–30° S), while a reduction of 2% yr^{-1} was recorded in northern hemisphere (30–50° N). The observed positive changes in O_3 concentration in southern hemisphere can be attributed to positive changes in TOC over southern South America, which is of higher magnitude than the observable negative changes over the oceans of southern hemisphere (Ebojie et al. 2016). Likewise, the negative O_3 trend of northern hemisphere can be explained by the stronger negative TOC changes over Europe and North America, which are more significant than the positive changes over northern China (Ebojie et al. 2016).

In tropics, O_3 concentrations showed distinct regional variations. Lelieveld et al. (2004) analyzed O_3 concentrations during 1977–2002 through in situ measurements in marine boundary layer and observed an increase of 0.4 ppb yr^{-1} for north tropical Atlantic. For tropical Pacific, a positive trend of 0.14 ppb yr^{-1} was recorded for Hawaii (Oltmans et al. 2013). Significant increases in tropospheric O_3 were recorded over some regions of South America (~2% yr^{-1}), South Asia (1–3% yr^{-1}), parts of Africa (~ 2% yr^{-1}) and over marine region (~2% yr^{-1}) (Ebojie et al. 2016). The positive trend of O_3 in these regions is attributed to the increase in population, industrialization and energy consumption (Cooper et al. 2014) along with meteorological changes and intercontinental transport of O_3/precursors over these regions (Parrish et al. 2012). Permadi and Oanh (2008) reported high surface O_3 levels in Jakarta during January 2002 to March 2004, which frequently exceeded hourly National Ambient Air Quality Standards (120 ppb). Wang et al. (2009) analyzed the variations in concentrations of tropospheric O_3 from 1994 to 2007 at a coastal site of Hong Kong and reported an increase of 0.86 ppb yr^{-1}.

4   Recent Trends in Ozone Concentration Over North America

Recent studies have shown that in North America surface O_3 has increased in eastern and arctic Canada, but no significant change in the O_3 concentration has been recorded over central and western Canada (Oltmans et al. 2013). Surface O_3 across the west coast of USA has increased significantly during the last decades (Parrish et al. 2012). The summertime O_3 decreased strongly in eastern USA, but has increased during winter season (Lefohn et al. 2010). O_3 concentrations are reported to be increasing over Mauna Loa in Hawaii, but have decreased over Miami Tori Shima in western North Pacific (Oltmans et al. 2012).

Parrish et al. (2009) studied the surface O_3 concentrations from several US coastal sites and analyzed baseline O_3 flowing onshore from North Pacific Ocean. This analysis showed significant increases in mean O_3 concentration in winter, spring and summer during 1980–2007 (Parrish et al. 2009). Cooper et al. (2010) studied the surface O_3 trends in free troposphere above western North America during springtime of 1984–2008, which is associated with increase in O_3 transport from atmospheric boundary layer of south and East Asia. Cooper et al. (2012) analyzed 21 year (1990–2010) rural and urban trends using all the data available from 12 to 41 sites in the western and eastern USA. The study showed that 43% sites of eastern USA and 17% sites of western USA showed a reducing trend in summertime O_3

concentration. Using 4th highest daily maximum 8 h average O_3 values (provided by USEPA), 71% of US sites showed significant decrease and 2% showed significant increase from 1980 to 2008, while for 1994–2008, O_3 concentration showed increments at 51 sites, while reduced significantly at 1% of the sites (Lefohn et al. 2010). Several studies related to photochemical processes suggest that reductions in the emissions of O_3 precursors are responsible for the decreasing trend of O_3 in eastern US (Butler et al. 2011; Hogrefe et al. 2011). Although the frequency of 8 h average O_3 shows a reducing trend, studies have confirmed a rise in baseline O_3 concentration (Lin et al. 2000). Cooper et al. (2012) also reported an increase of 0.41 ppbv yr^{-1} in the springtime baseline O_3 during the period 1995–2011. Many studies have provided sufficient evidences that Asian pollution plumes are responsible for increasing baseline O_3 concentration (Jacob et al. 1999; Dentener et al. 2011) with strongest implications for western US (Lin et al. 2015; Verstraeten et al. 2015).

5 Recent Trends in Ozone Concentration Over Europe

Between 1950s and 2000s, O_3 concentration in Europe has probably doubled, however, trends have shown decrease in the concentration of O_3 over Western Europe since 2000 (Hartmann et al. 2013). Lemaire et al. (2016) used statistical models to study the O_3 concentration across Europe and to analyze the patterns of increase in O_3 concentration by the end of the century (2071–2100). This study divided the European continent into eight zones and O_3 concentration over these zones was determined. The results showed that average daily O_3 concentration reached its maximum value during the months of June, July and August between 1976 and 2005 (Lemaire et al. 2016). A north south gradient of O_3 concentration was observed with lower concentration in North and higher in the Mediterranean region. The average O_3 predictions for summers at the end of the century (2071–2100) also showed a similar pattern with high O_3 concentration observed in the southern part of the selected transect. This study showed that O_3 concentration increased by 5.3, 5.9, 5.8, 5.0 and 2.4 ppb, respectively in Eastern Europe, France, Iberian Peninsula, mid Europe and northern Italy. Lemaire et al. (2016) established a significant increase in O_3 concentration over Eastern Europe, Mediterranean land surface and North Africa and recorded a decrease over British Irish Isles and Scandinavia. Hartmann et al. (2013) also reported high concentration of O_3 in southern Europe due to favourable climatological conditions for O_3 formation.

According to a survey conducted at different sites of over seventeen countries of European Union (EU), O_3 levels exceeded the EU target value for protecting human health for about 25 times in 2011 at 40, 24, 21 and 9% of rural, suburban, industrial and traffic sites (Guerreiro et al. 2014). An analysis of 8 h mean O_3 concentration over different sites of E-27 countries in Europe observed a significant decreasing trend over 18% of the selected stations, while 2% (mostly in the Iberian Peninsula) registered a significant increasing trend for a period between 2002 and 2012 (Guerreiro et al. 2014).

6   Recent Trends in Ozone Concentration Over Asia

Air pollution in east and south Asia has gathered increasing attention from the scientific community and policy makers in the last few decades. These areas not only experience high concentration of O_3, but are also an important source of O_3 precursors, which as already discussed play a significant role in intercontinental O_3 transport. (Wang 2005) selected sixteen sites in East Asia and analyzed the O_3 concentration at each site by using Acid Deposition Monitoring Network in East Asia (EANET) during 2000–2004. The study included ten sites from Japan, three from Republic of Korea, one from Russia and two from Thailand. A close analysis of the data suggested the sites over mid latitudes had high O_3 concentration and the tropical sites recorded the lowest concentration of O_3. A few mid latitude sites like Mondy in Russia, Kanghwa in Korea and Happo in Japan had annual average mean O_3 concentration of 44 ppb (3 yearly average), 41 ppb (3 yearly average) and 55 ppb (5 yearly average), respectively, while both the tropical sites of Thailand recorded two yearly average mean concentration of O_3 to be 20 ppb (Wang 2005). The high concentration of O_3 at the mid latitudes can be correlated to the frequent occurrence of cyclones during the spring season which transports O_3 from high latitudes. Correlation studies showed a positive correlation between O_3 concentration and wind speed at Mondy (Russia), indicating a significant role of EURO-ASIA O_3 transport in maintaining a high O_3 level at this site. However, at Happo (Japan) a negative correlation was recorded with humidity, which suggests that sinking of dry air from the stratosphere may also play a noteworthy role in enhancing O_3 concentration (Wang 2005). Verstraeten et al. (2015) studied the recent trends of tropospheric O_3 over China by using TM5 global chemistry model, which predicted a 7% increase in O_3 concentration over China during 2005–2010. This increase is attributed mainly to rise in emissions of O_3 precursors and downward transport of stratospheric O_3. Measurements done by Tropospheric emission spectrometer (TES) recorded 1.08% yr^{-1} increase in O_3 concentration over eastern China from 2005 to 2010. TM5 simulation studies suggested that stratospheric tropospheric transport (STE) contributed 0.8% yr^{-1} (0.78–0.84% yr^{-1}) and NOx emissions 0.58% yr^{-1}, towards the observed increment rate of tropospheric O_3 over China (Verstraeten et al. 2015). Sun et al. (2016) observed that during 2003–2015, an increase of 2.1 ± 0.9 ppb yr^{-1} was recorded in summertime O_3 concentration at Mt. Tai in central China.

Wang et al. (2014) examined the temporal and spatial variations of major pollutants in 31 provincial capital cities of China, based upon the data for the period between March 2013 and February 2014. The results clearly indicated that air quality of 24 out of 31 cities did not fulfill the Chinese Ambient Air Quality Standards (CAAQS). Eight hourly O_3 concentration varied from 23 to 51 ppb in North China, 29–53 ppb in South East China and 32–50 ppb in West China (Wang et al. 2014). On the basis of Air Quality Index (AQI) system, O_3 was identified as the most important pollutant at most of the selected sites. During summer, O_3 was the most frequent pollutant in South East and West regions and the second most frequent (after particulate matter) in the North region (Wang et al., 2014). The percent contribution of O_3 to total pollution was 33.1, 75.6 and 54.8%, respectively in North, South East

and West China during summer season (Wang et al. 2014). Lee et al. (2015) analyzed summertime O_3 concentration over East Asia using Integrated Climate and
Air Quality Modelling system (ICAMS) for the next three periods i.e. 2016–2025
(2020s), 2046–2055 (2050s) and 2091–2100 (2090s). Based on this study, annual
daily mean 8 h surface O_3 concentration for summertime was predicted to change in
the range of 2–8 ppb, −3 to 8 ppb and −7 to 9 ppb for 2020s, 2050s and 2090s,
respectively (Lee et al. 2015). These changes were mainly attributed to the changes
in NOx and NMVOC emissions for the developing countries like China, whereas in
case of developed countries like Japan, the projected O_3 concentrations were more
influenced by regional climate change than increase in precursor emissions (Lee
et al. 2015).

Li et al. (2016) investigated daily mixing ratios at different sites in East Asia using
a chemical transport model including an online tracer tagged procedure, which gives
a clear indication of the source of O_3 present at a particular site. Through this study, it
was observed that long range transport from outside East Asia contributed to the greatest fraction to annual surface O_3 over remote regions, the Korean peninsula and Japan
reaching 50–80%, while in China, it was 5–10% of the total O_3. An analysis of daily
surface O_3 showed distinct patterns that varied at different sites. On Asian continent
and in Japan, daily surface O_3 showed a single peak (30–50 ppbv) distribution, while
at low latitude sites away from the continent, two peaks at 10–20 ppbv and 40–50
ppbv were recorded (Li et al. 2016). This discrepancy in O_3 behaviour can be explained
by the continental and oceanic air masses prevalent in this region (Li et al. 2016).

As India is a developing country, the industrialization and urbanization processes
act as important sources of emission of O_3 precursors. The tropical climatic conditions prevalent in India also provide favourable conditions for O_3 formation (Tiwari
et al. 2008). Selected O_3 monitoring data are available for different parts of the
country. Pandey et al. (1992) recoded 24 h annual mean O_3 concentrations ranging
between 6.12 and 10.2 ppb during 1989–1990 around Varanasi. During the same
period, 9 h mean daytime O_3 concentrations in Delhi varied from 9.4 to 128.31 ppb
(Varshney and Aggarwal 1992). Singh et al. (1997) observed that 10 h ground level
mean O_3 concentrations in Delhi varied between 34 and 126 ppb during winter season in 1993. An annual average daytime O_3 concentration of 27 ppb at Pune was
reported during August 1991 to July 1992 (Khemani et al. 1995). Lal et al. (2000)
studied the patterns of O_3 concentrations from 1991 to 1995 at an urban site at
Ahmadabad and reported that daytime mean O_3 concentrations exceeding 80 ppb
was rarely observed. Table 1.1 shows the monitoring data recorded from different
monitoring sites across India.

Jain et al. (2005) studied extensively the variations in O_3 concentrations over
Delhi, which experiences intense anthropogenic activities leading to a high rate of
emissions of O_3 precursors. The monthly average maximum concentrations ranged
between 62 and 95 ppb in summer, whereas in autumn, it was found to be 50–82 ppb
(Jain et al. 2005). The study also showed that for several days, surface O_3 values at
Delhi exceeded the World Health Organization (WHO) ambient air quality standard (hourly average of 80 ppb) (Jain et al. 2005). Kulkarni et al. (2010) studied
the changes in tropospheric O_3 over Delhi, Bangalore and Hyderabad during

Table 1.1 Variations in ozone concentrations recorded at different monitoring sites in India

S. no	Site/Location	Time of the experiment	Exposure period	Concentration (ppb)	References
1.	New Delhi	1989–1990	9 h	9.4–128.3	Varshney and Aggarwal (1992)
		08/1991–07/1992	Annual average	27	Khemani et al. (1995)
		1993	10 h	34–126	Singh et al., (1996)
		1997–2003	Monthly average	50–95	Jain et al. (2005)
2.	Varanasi	02/2002–02/2002	8 h	33–40	Mittal et al. (2007)
		09/2002–02/2006	8 h	24–62.35	Tiwari et al. (2008)
3.	Durgapur	02/2013–05/2013	–	68	Dey et al. (2014)
4.	Anantpur	2001–2003	Annual average	35.9	Ahammed et al. (2006)
		01/2010–12/2010	-do-	40.7	Reddy et al. (2013)
5.	Gadanki	11/1993–12/1996	-do-	34	Naja and Lal (2002)
6.	Kannur	11/2009–12/2010	-do-	18.4	Nishanth et al. (2014)
7.	Thumba	04/1997–03/1998	-do-	23	Nair et al. (2002)
8.	Trivandrum	11/2007–05/2009	-do-	11.5–2.1	David and Nair (2011)
9.	Nanital	10/2006–12/2008	-do-	43.9	Kumar et al. (2010)
10.	Mt Abu	01/1993–12/2006	-do-	39.9	Naja et al. (2003)
11.	Nagercoil	2007–2010	-do-	18.9–20.0	Elampari and Chithambarathanu (2011)
12.	Kanpur	06/2009–05/2013	-do-	27.9	Gaur et al. (2014)
13.	Agra	01/2002–12/2002	-do-	21	Satsangi et al. (2004)
14.	Pune	01/2001–12/2005	-do-	30.1	Debaje and Kakade (2009)
		06/2003–05/2004	-do-	30.9	Beig and Singh (2007)
15.	Ahemadabad	06/2003–05/2004	-do-	20.7	Beig and Singh (2007)

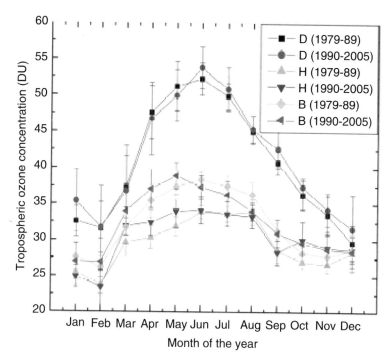

Fig. 1.4 Annual variations in average Tropospheric Ozone over the three major cities, Delhi, Hyderabad and Bangalore in India during 1979–1989 and 1990–2005 (Source: Kulkarni et al. 2010)

1990–2005, as compared to 1979–1989 using the Tropospheric Ozone Residual (TOR) data and observed that O_3 values over the three cities increased significantly during 1990–2005 as compared to 1979–1989 (Fig. 1.4). A detailed analysis of O_3 concentrations showed a different trend in Delhi as compared to Hyderabad and Bangalore. TOR data suggested an increase in O_3 concentration over Delhi during monsoon, post monsoon and winter months, whereas no increase was recorded during the pre-monsoon months. Hyderabad and Bangalore showed somewhat similar trend with increments in O_3 concentration during pre monsoon and post monsoon months with no increase in winter months. However, data suggested a decrease in O_3 concentration during monsoon months in Hyderabad, whereas no significant change was observed in Bangalore (Kulkarni et al. 2010).

Dey et al. (2014) studied O_3 concentrations from February 2013 to May 2013 and reported O_3 concentrations as high as 66.8 ppb at Durgapur, West Bengal. Ganguly (2012) analyzed O_3 concentrations between 1998 and 2008 at five O_3 monitoring stations, namely New Delhi, Nagpur, Pune, Kodiakanal and Thiruvanathpuram, stretching across the country. During the study period, the surface O_3 levels at Pune and Thiruvanantpuram indicated a small decreasing trend, while at New Delhi, Nagpur and Kodaikanal, an increasing trend was observed (Fig. 1.5). Ganguly (2012) calculated the upper control limit (UCL) for surface O_3 at different Indian

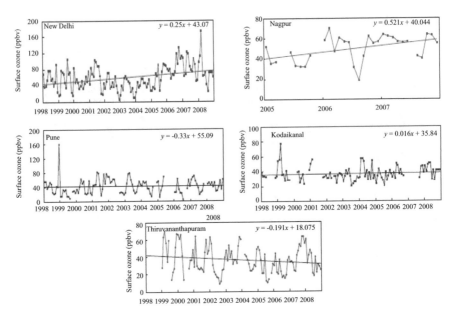

Fig. 1.5 Surface ozone trends at a few selected cities in India from 1998 to 2008 (Source: Ganguly 2012)

stations under study as described in Gupta and Kapoor (1993). The UCL was found to be 79.5, 57.6, 57.3, 43.4 and 47.3 ppb for New Delhi, Nagpur, Pune, Kodiakanal and Thiruvanathpuram, respectively. When the observed value of O_3 exceeded the UCL, it indicated some new sources of variations, be it natural or anthropogenic (Ganguly 2012).The study of Ganguly (2012) also emphasized the role of stratosphere- troposphere exchange (STE) apart from transport processes and photochemical production, in influencing the concentrations of O_3 in the troposphere and estimated that STE influences the surface O_3 levels in the Indian cities by 8–16%.

Mittal et al. (2007) using regional episodic chemical transport model (HANK) predicted 8 h daily O_3 concentrations varying in a range 33–40 ppb in Varanasi during the period February–April 2000. Debaje and Kakade (2008) also reported similar range of O_3 concentrations from urban and rural sites of Maharashtra. Regional Chemistry Transport Model (REMO-CTM) showed high AOT40 values, especially between November and May that exceeded the threshold set by WHO for agricultural crops (3 ppmh for 3 months) (Roy et al. 2009).

Data from continuous monitoring studies done at a suburban site in Varanasi, India during the period 2002–2012 not only showed a significant increase in O_3 concentrations, but also recorded well defined seasonal variations during the study period (Fig. 1.6). The average monthly O_3 concentration was found to exceed frequently the National Ambient Air Quality Standards (NAAQS) as set by the Central Pollution Control Board (CPCB), New Delhi. The experimental site experienced highest summertime O_3 concentration followed by winter and rainy seasons (Tiwari et al., 2008). The summertime monthly O_3 concentration varied from 55.21 ppb to 62.2 ppb during the study period (Singh et al. 2014; Sarkar et al. 2015), whereas in

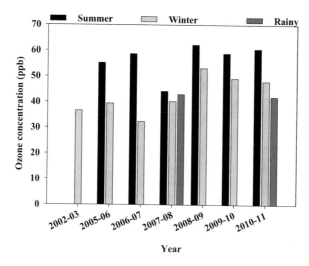

Fig. 1.6 Variations in ozone concentrations in Varanasi from 2002 to 2012

winter season, it varied from 32.33 ppb to 53.2 ppb (Mishra et al., 2013) (Fig. 1.6). The variation in rainy season during the study period could not be evaluated due the lack of sufficient monitoring data.

7   Seasonal and Diurnal Variations in Ozone Concentrations

Prominent seasonal variations in O_3 concentrations in the troposphere are well documented (Tiwari et al. 2008; Gaur et al. 2014; Xu et al. 2016). The factors determining the tropospheric O_3 concentrations viz. the rate of emission of O_3 precursors, existing meteorological conditions and Stratospheric- Tropospheric Exchange, play significant roles in bringing about seasonal and diurnal variations. O_3 monitoring studies conducted at a suburban site in Varanasi from 2002 to 2006 (Tiwari et al. 2008) and at an urban site in Kanpur from 2009 to 2013 (Gaur et al. 2014), clearly showed sharp seasonal and diurnal variations in O_3 concentrations. The hourly maxima were observed between 1200 and 1400 h, while evening and morning hours showed lower O_3 concentrations (Tiwari et al. 2008). Similar seasonal variation trends were also recorded by Gaur et al. (2014) with peak concentrations during summer and least during monsoon season (Fig. 1.7).

The rise in O_3 concentrations during summer season can be attributed to its linear relationship with intense solar radiation, which directly influences the chemical kinetic rates and mechanism pathways for O_3 production (Han et al. 2011; Pudasainee et al. 2006). The dependence of O_3 concentrations on solar radiations is evident by the observed positive correlation between the two parameters (Tiwari et al. 2008). The low O_3 levels observed during rainy/monsoon season result due to washout effects and reduced availability of precursors for production of O_3. Higher O_3 concentrations during winter (as compared to monsoon season) are a result of long range

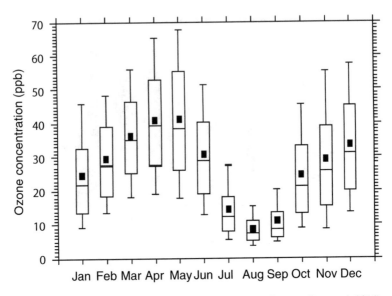

Fig. 1.7 Seasonal variations in ozone concentrations at Kanpur (Source: Gaur et al. 2014)

transport. Further, low temperature during winters decreases the thermal decomposition of O_3 precursors, thus increasing their atmospheric lifetime. (Gaur et al. 2014).

Lal et al. (2014) studied average seasonal profiles of O_3 mixing ratio in the troposphere using balloon borne electrochemical concentration cell (ECC) sensors above Ahmedabad from May 2003 to July 2007. During winter months of December, January and February, average O_3 concentrations were about 28 ppbv near the surface, 56 ppbv near 1 km and about 45 ppbv at 3.5 km. Near surface O_3 values were lowest during monsoon (~18 ppbv) season, which include the months of June, July and August (Lal et al. 2014). In middle troposphere, average O_3 was maximum in summer and monsoon (~ 66 ppbv) followed by spring and autumn. The average O_3 in the upper troposphere was highest in spring, particularly in March and April, and lowest in autumn (September, October and November). This study suggested that highest O_3 values in free troposphere above Ahmedabad were recorded during spring and lowest during autumn (Lal et al. 2014). The tropospheric O_3 column (TOC) was greatest in April (~44 DU) and least (~34–35 DU) during July, August and September (Lal et al. 2014). Agrawal et al. (2003) reported daytime 6 h mean O_3 concentrations at Varanasi as 15.44 and 55.70, during winter and summer seasons, respectively. Seasonal and diurnal variations have also been recorded at a tropical site in India (Nagercoil) for a period from 2007 to 2010 (Elampari and Chithambarathanu 2011). A gas sensitive semiconductor (GSS) sensor based O_3 monitor was used to measure the surface O_3 concentrations and the data revealed that O_3 is a function of time and varies from morning minimum to afternoon maximum. In addition to this, distinct seasonal variations were also recorded with highest O_3 concentrations observed during summers (38.45–10.31 ppb), followed by south west monsoon (33.44–8.86 ppb) and winters (33.30–10.55 ppb) (Elampari and Chithambarathanu 2011).

Seasonal variations in O_3 concentrations have also been reported in several other studies conducted in USA (Shena and Mickley 2017; Vingarzan 2004), Europe (Varotsos et al. 2001, 2003), Canada (Monks 2000), Korea (Ghim and Chang 2000; Kim et al. 2006) and Portugal (Kulkarni et al. 2016). Ismail et al. (2011) analyzed the O_3 variations by studying monthly observations (144) from January 1996 to December 2007 in Kemaman, a developing town in Malasiya. ARIMA (Box Jenkins Autoregressive Integrated Moving Average) model was applied to O_3 data to calculate the seasonal index (SI). SI depicted a well defined annual cycle with highest mean O_3 concentration occurring in August (122.058) and lowest in November (80.047) (Ismail et al. 2011). Seasonal variation pattern observed in this study differed from other countries such as USA, UK, Italy, Canada, Japan, etc. as the peak O_3 concentrations did not correspond to maximum photochemical activity in summer. Wang et al. (2014) analyzed 1 year O_3 data (March13–February 14) at 31 sites in China and observed prominent seasonal variations in 8 h O_3 concentration with highest concentration in summer season and lowest in winter. This variation can be attributed to the fact that O_3 formation rate largely depends upon selected meteorological conditions such as intensity of solar radiation (Chan and Chan 2000; Atkinson 2000).

Xu et al. (2016) studied 20 year (1994–2013) surface O_3 mixing ratio at Mt. Walignam site in the north eastern Tibetan plateau region in China by analyzing measurements of Global Atmospheric Watch (GAW) programme by World Meteorological Organization (WMO) and recorded prominent seasonal variations with maximum O_3 concentration recorded during summer and minimum in winter. In this study, the average seasonal peak was observed during June–July (Xu et al. 2016). Several studies have recorded similar trends in seasonal fluctuations of O_3 concentration and have attributed this behavior to different causes eg frequent STE events (Tang et al. 2011), enhanced vertical convection (Ma et al. 2014), long range transport from eastern central China, central southern Asia or Europe during summer (Zhu et al. 2004) and strong cross country transport and vertical convection during the east Asian summer monsoon (Yang et al. 2014). However, Zheng et al. (2011) suggested that since STE reaches maximum strength in spring and start declining in late spring, increased summertime O_3 concentration is mainly attributed to photochemical production. Hayashida et al. (2015) also observed high O_3 values in summer months and low during winter over central and eastern China. MOZAIC (Measurements of O_3 and water vapour by Airbus In-Service Aircraft) measurements during 2004–2005 showed that above 6 km altitude, the O_3 mixing ratio was as high as 80–100 ppbv during July 2005. During a nine year study period between 2005 and 2013, prominent seasonal variations, with high O_3 concentrations during summer were recorded by OMI (Ozone Monitoring Instrument) measurements (Hayashida et al. 2015). Enhancement in O_3 concentration in upper troposphere is mainly attributed to stratospheric O_3 intrusion while O_3 mixing ratio in lower troposphere was due to photochemical production.

Prominent daily variation is also a distinguishing feature of O_3 pollution which is reported through several studies. Tiwari et al. (2008) and Awang et al. (2015) observed high daytime and low nighttime O_3 concentrations at Varanasi, India and Kemaman, Malasiya, respectively. This trend of daily variation can be explained due to high temperature and high concentration of O_3 precursors (mainly NOx) during the daytime as compared to nighttime (Tiwari et al. 2008). Xu et al. (2016) also observed daily variations in O_3 mixing ratio at Mt. Waliguan GAW station, China, which showed a reverse trend with high nighttime and low daytime O_3 concentration. This unusual trend of O_3 concentration can be explained by the geographical location of the experimental site, which experiences upslope winds influenced by boundary layer during the daytime and downslope winds influenced by free tropospheric air during night. The boundary layer air masses are characterized by lower O_3 mixing ratios in comparison with the free tropospheric air masses, which have high O_3 concentration. It was further observed that occurrence time of daily minimum O_3 mixing ratio showed a distinct annual variation (Xu et al. 2016). However, O_3 minimum occurred in June and August, later than expected, which suggested that photochemical production which is maximum at early noon during summertime, rather than boundary layer movement is the major cause of this variation (Xu et al. 2016). The phenomena, however, requires further investigation. The largest nighttime increasing rate of O_3 mixing ratio occurred in autumn (0.29 ± 0.11 ppbv yr^{-1}), followed by spring (0.24 ± 0.12 ppbv yr^{-1}), summer (0.22 ± 0.20 ppbv yr^{-1}) and winter (0.13 ± 0.10 ppbv yr^{-1}) (Xu et al. 2016).

8   Future Predictions in Ozone Concentrations

The background O_3 concentration is further expected to increase at an annual rate of 0.3 ppb yr^{-1} (Wilkinson et al. 2012). Under current emission rates and legislation scenarios, future surface O_3 may show a decrease of 2 ppb in cleaner areas or an increase of 4 ppb in polluted areas between 2000 and 2030 (Dentener et al. 2006). Global future tropospheric O_3 concentrations are modeled using Global Chemical Transport Models driven by future climate and meteorological scenarios and estimates of future emission. Dentener et al. (2006) assessed the changes in surface O_3 between 2000 and 2030 using 26 global atmospheric chemistry models and three different emission scenarios, viz. CLE scenario (reflects implementation of current air quality legislation), MRC scenario (represents the maximum emission reduction of O_3 precursors), and an intermediate IPCC SRES A2 scenario. By 2030, global O_3 concentration is expected to increase by 1.5 ± 1.2 ppb under CLE scenario and 4.3 ± 2.2 ppb under A2 scenario (Dentener et al. 2006). Asian regions are predicted to experience maximum increase in surface O_3, while Europe and North America are expected to see moderate increase (Dentener et al. 2006). Meehl et al. (2007) have estimated an increase of 20–25% O_3 concentrations by 2050 and 40–60% by 2100.

9   Conclusion

The chapter examines the long term assessment of O_3 trends around different parts of the globe. The continuously increasing trend of O_3 has placed this gaseous pollutant as the most important agenda in the priority list that needs immediate solution. The significant role that the meteorological factors play in the formation of O_3 adds up to the already existing problem, in view of the climate change scenario, which tends to create conditions favourable for O_3 formation. Although, a few regions of mid latitude have shown a decline in O_3 trends owing to the pollution control strategies, which have reduced the emission of O_3 precursors, the overall O_3 budget shows an alarming increment. Regions like North America and Europe have shown a decline in regional O_3 concentration, which may be associated with reduction in NOx and other tropospheric O_3 precursor emissions over these regions. However, due to long range transboundary O_3 transportation of O_3 precursors from South and East Asia, the regional reduction in O_3 concentration in developed countries of North America becomes insignificant. The transboundary transportation issue emphasizes the importance of O_3 at hemispheric scale. Regions of Asia, particularly East and South Asia have presently emerged as hotspot regions for O_3 formation. In addition to the increasing trends of O_3 formation, several studies have also confirmed prominent seasonal and diurnal variations in O_3 concentration over different regions, which can be attributed to the differences in the prevailing meteorological conditions. Satellite instruments have quantified the present day tropospheric O_3 burden on global basis and have observed a significant increase in tropospheric O_3 column over extended tropical regions of southern Asia as well as regions of south and north Pacific Ocean. However, changes observed in the tropospheric O_3 column through various modeling studies have shown small variations as compared with the results obtained from the monitoring studies of the tropospheric O_3 trends in the last few years. This observation can be explained by the slowdown in the growth rate of O_3 precursors. The tedious implementation of the legislations on air pollution related policies contributed to the reduction in tropospheric O_3 levels and its precursor emissions. As the O_3 levels in the troposphere are likely to change in near future, there remains a need for continuing research to monitor the O_3 concentrations in different regions and to evaluate its impacts and interaction with the ecosystem.

References

Agrawal M, Singh B, Rajput M, Marshall F, Bell JNB (2003) Effect of air pollution on periurban agriculture: a case study. Environ Pollut 126:323–329

Ahammed YN, Reddy RR, Gopal KR, Narasimhulu K, Basha DB, Reddy LSS, Rao TVR (2006) Seasonal variation of the surface ozone and its precursor gases during 2001–2003, measured at Anantapur (14.62°N), a semi-arid site in India. Atmos Res 80:151–164. https://doi.org/10.1016/j.atmosres.2005.07.002

Anfossi D, Sandroni S, Viarengo S (1991) Tropospheric ozone in the nineteenth century: the Moncalieri series. J Geophys Res 96:17349–17352

Atkinson R (2000) Atmospheric chemistry of VOCs and NOx. Atmos Environ 34:2063–2101

Awang NR, Ramli NA, Yahaya AS, Elbayoumi M (2015) High nighttime ground-level ozone concentrations in kemaman: NO and NO_2 concentrations attributions. Aerosol Air Qual Res 15:1357–1366

Baier M, Kandlbinder A, Golidack D, Dietz K-J (2005) Oxidative stress and ozone: perception, signalling and response. Plant Cell Environ 28:1012–1020

Beig G, Singh V (2007) Trends in tropical tropospheric column ozone from satellite data and MOZART model. Geophys Res Lett 34:L17801

Bojkov RD (1986) Surface ozone during the second half of the nineteenth century. J Clim Appl Meteorol 25:343–352

Brown-Steiner B, Hess P (2011) Asian influence on surface ozone in the United States: a comparison of chemistry, seasonality, and transport mechanisms. J Geophys Res 116:D17309

Butler TJ, Vermeylen FM, Rury M, Likens GE, Lee B et al (2011) Response of ozone and nitrate to stationary source NOx emission reductions in the eastern USA. Atmos Environ 45:1084–1094

Cartalis C, Varotsos C (1994) Surface ozone in Athens, Greece, at the beginning and at the end of the twentieth century. Atmos Environ 28:3–8

Chameides W, Walker JCG (1973) A photochemical theory of tropospheric ozone. J Geophys Res 78(36):8751–8760. https://doi.org/10.1029/JC078i036p08751

Chan CY, Chan LY (2000) The effect of meteorology and air pollutant transport on ozone episodes at a subtropical coastal asian city, Hong Kong. J Geophys Res 105:20707–20719

Cooper O, Ziemke J (2013) Tropospheric ozone [in "State of the Climate in 2012"]. Bull Am Meteorol Soc 94(8):S38–S39

Cooper OR, Parrish DD, Stohl A, Trainer M, Nédélec P et al (2010) Increasing springtime ozone mixing ratios in the free troposphere over western North America. Nature 463(7279):344–348. https://doi.org/10.1038/nature08708

Cooper OR, Gao R-S, Tarasick D, Leblanc T, Sweeney C (2012) Long-term ozone trends at rural ozone monitoring sites across the United States, 1990–2010. J Geophys Res 117:D22307. https://doi.org/10.1029/2012JD018261

Cooper OR et al (2014) Global distribution and trends of tropospheric ozone: an observation-based review. Elem Sci Anth 2:000029

Crutzen PJ (1973) A discussion of the chemistry of the stratosphere and troposphere. Pure Appl Geophys 106–108:1385–1399

Crutzen PJ (1974) Photochemical reactions initiated by and influencing ozone in the unpolluted troposphere. Tellus 26:47–57

David LM, Nair PR (2011) Diurnal and seasonal variability of surface ozone and NOx at a tropical coastal site: association with mesoscale and synoptic meteorological conditions. J Geophys Res 116(D10):D10303

Debaje SB, Kakade AD (2008) Surface ozone variability over western Maharashtra, India. J Hazard Mater 161:686–700

Debaje SB, Kakade AD (2009) Surface ozone variability over western Maharashtra, India. J Hazard Mater 161(2–3):686–700

Dentener F, Kinne S, Bond T, Boucher O, Cofala J et al (2006) Emissions of primary aerosol and precursor gases in the years 2000 and 1750 prescribed datasets for AeroCom. Atmos Chem Phys 6(4):321–4344

Dentener F, Keating T, Akimoto H (eds) (2011) Hemispheric transport of air pollution 2010: part A: ozone and particulate matter. UN, New York. (Air Pollut. Stud, vol. 17)

Derwent RG, Collins WJ, Johnson CE, Stevenson DS (2001) Transient behavior of tropospheric ozone precursors in a global 3D CTM and their indirect greenhouse effects. Clim Chang 49:463–487

Dey S, Pati C, Gupta S (2014) Measurement and analysis of surface ozone and its precursors at three different sites in an urban region in eastern India. Environ Forensic 2014:112–120

Distribution of ozone in the lower troposphere over the Bay of Bengal and the Arabian Sea during ICARB (2006) Effects of continental outflow. J Geophys Res https://doi.org/10.1029/2010JD015298

Doherty RM (2015) Atmospheric chemistry: ozone pollution from near and far. Nat Geosci 8:664–665

Donner LJ, Wyman BL, Hemler RS, Horowitz LW, Ming Y et al (2011) The dynamical core, physical parameterizations, and basic simulation characteristics of the atmospheric component of the GFDL global coupled model CM3. J Clim 24(3):484–3,519. https://doi.org/10.1175/2011JCLI3955

Duenas C, Fernández MC, Cañete S, Carretero J, Liger E (2005) Stochastic model to forecast ground-level ozone concentration at urban and rural areas. Chemosphere 61:1379–1389

Ebojie F, Burrows JP, Gebhardt C, Ladstätter-Weißenmayer A, von Savigny C, Rozanov A, Weber M, Bovensmann H (2016) Global tropospheric ozone variations from 2003 to 2011 as seen by SCIAMACHY. Atmos Chem Phys 16:417–436

Elampari K, Chithambarathanu T (2011) Diurnal and seasonal variations in surface ozone levels at tropical semi-urban site, Nagercoil, India, and relationships with meteorological conditions. Int J Sci Technol 1(2):80–88

Ellingsen K, Gauss M, Van Dingenen R, Dentener FJ, Emberson L, Fiore AM, Schultz MG, Stevenson DS, Ashmore MR, Atherton CS, Bergmann DJ, Bey I, Butler T, Drevet J, Eskes H, Hauglustaine DA, Isaksen IAS, Horowitz LW, Krol M, Lamarque JF, Lawrence MG, van Noije T, Pyle J, Rast S, Rodriguez J, Savage N, Strahan S, Sudo K, Szopa S, Wild O (2008) Global ozone and air quality: a multi-model assessment of risks to human health and crops. Atmos Chem Phys Discuss 8:2163–2223

Emberson LD, Buker P, Ashmore MR, Mills G, Jackson LS, Agrawal M, Atikuzzaman MD, Cinderby S, Engardt M, Jamir C, Kobayashi K, Oanh NTK, Quadir QF, Wahid A (2009) A comparison of North American and Asian exposure–response data for ozone effects on crop yields. Atmos Environ 43:1945–1953

Fiore AJ, Obreman M-Y, Lin L, Zhang O, Clifton D, Jacob V, Naik L, Horowitz J, Milly G (2014) Estimating North American background ozone in U.S. surface air with two independent global model variability, uncertainties, and recommendations. Atmos Environ 96:284–300

Fischer EV, Jaffe DA, Reidmiller DR, Jaeglé L (2010) Meteorological controls on observed peroxyacetyl nitrate at Mount Bachelor during the spring of 2008. J Geophys Res 115:D03302

Fowler D, Pilegaard K, Sutton MA, Ambus P, Raivonen M, Duyzer J, Simpson D, Fagerli H, Fuzzi S, Schjoerring JK, Granier C, Neftel A, Isaksen ISA, Laj P, Maione M, Monks PS, Burkhardt J, Daemmgen U, Neirynck J, Personne E, Wichink-Kruit R, Butterbach-Bahl K, Flechard C, Tuovinen JP, Coyle M, Gerosa G, Loubet B, Altimir N, Gruenhage L, Ammann C, Cieslik S, Paoletti E, Mikkelsen TN, Ro-Poulsen H, Cellier P, Cape JN, Horvath L, Loreto F, Niinemets U, Palmer PI, Rinne J, Misztal P, Nemitz E, Nilsson D, Pryor S, Gallagher MW, Vesala T, Skiba U, Brueggemann N, Zechmeister-Boltenstern S, Williams J, O'Dowd C, Facchini MC, de Leeuw G, Flossman A, Chaumerliac N, Erisman JW (2009) Atmospheric composition change: ecosystems-atmosphere interactions. Atmos Environ 43:5193–5267. https://doi.org/10.1016/j.atmosenv.2009.07.068

Ganguly ND (2012) Influence of stratospheric intrusion on the surface ozone levels in India. ISRN Meteorol 2012 1:7. Article ID 625318

Gaur A, Tripathi SN, Kanawade VP, Tare V, Shukla SP (2014) Four-year measurements of trace gases (SO_2, NO_x, CO, and O3) at an urban location, Kanpur, in Northern India. J Atmos Chem 71:283–301

Ghim YS, Chang YS (2000) Characteristics of ground-level ozone distribution on Korea for the period of 1990- 1995. J Geophys Res 105:8877–8890

Glotfelty T, Zhang Y, Karamchandani P, Streets DG (2014) Will the role of intercontinental transport change in a changing climate? Atmos Chem Phys 14:9379–9402. https://doi.org/10.5194/acp-14-9379-2014

Granier C, Bessagnet B, Bond T, D'Angiola A, van der Gon HD et al (2011) Evolution of anthropogenic and biomass burning emissions of air pollutants at global and regional scales during the 1980–2010 period. Clim Chang 109:163–190. https://doi.org/10.1007/s10584-011-0154-1

Guerreiro CBB, Foltescu V, de Leeuw F (2014) Air quality status and trends in Europe. Atmos Environ 98:376–384

Gupta SC, Kapoor VK (1993) Fundamentals of applied statistics, chapter 1. Sultan Chand and Sons, New Delhi

Haagen-Smit AJ (1952) Chemistry and physiology of Los Angeles smog. Ind Eng Chem 44:1342

Hayashida S, Liu X, Ono A, Yang K, Chance K (2015) Observation of ozone enhancement in the lower troposphere over East Asia from a space-borne ultraviolet spectrometer. Atmos Chem Phys 15:9865–9881

Han S, Bian H, Feng Y, Liu A, Li X, Zeng F, Zhang X (2011) Analysis of the relationship between O_3, NO and NO_2 in Tianjin, China. Aerosol Air Qual Res 11:128–139

Hartmann DL, Klein Tank AMG, Rusticucci M, Alexander L, Brönnimann S et al (2013) Observations: atmosphere and surface supplementary material. In: Stocker TF, Qin D, Plattner G-K, Tignor M, Allen SK (eds) Climate change 2013: the physical science basis. Contribution of working group I to the fifth assessment report of the intergovernmental panel on climate change. Cambridge University Press, New York. Available from www.climatechange2013.org and www.ipcc.ch

Hogrefe C et al (2011) An analysis of long-term regional-scale ozone simulations over the Northeastern United States: variability and trends. Atmos Chem Phys 11:567–582. https://doi.org/10.5194/acp-11-567-2011

Horowitz LW, Walters S, Mauzerall DL, Emmons LK, Rasch PJ et al (2003) A global simulation of tropospheric ozone and related tracers: description and evaluation of MOZART, version 2. J Geophys Res 108:4784. https://doi.org/10.1029/2002JD002853,D24

HTAP (2010) Hemispheric transport of air pollution. UNECE, Geneva

Huang M et al (2010) Impacts of transported background ozone on California air quality during the ARCTAS-CARB period–a multi-scale modeling study. Atmos Chem Phys 10:6947–6968. https://doi.org/10.5194/acp-10-6947-2010

IPCC (Intergovernmental Panel on Climate Change) (2013) Working Group I contribution to the IPCC fifth assessment report "Climate change 2013: the physical science basis", Final Draft Underlying Scientific-Technical Assessment. Available at http://www.ipcc.ch

Ismail M, Ibrahim MZ, Ibrahim ATG, Abdullah AM (2011) Time series analysis of surface ozone monitoring records in Kemaman, Malaysia. Sains Malays 40(5):411–417

Jacob DJ, Logan JA, Murti PP (1999) Effect of rising Asian emissions on surface ozone in the United States. Geophys Res Lett 26:2175–2178. https://doi.org/10.1029/1999GL900450

Jaffe D et al (1999) Transport of Asian air pollution to North America. Geophys Res Lett 26:711–714. https://doi.org/10.1029/1999GL900100

Jain SL, Arya BC, Kumar A, Ghude SD, Kulkarni PS (2005) Observational study of surface ozone at New Delhi, India. Int J Remote Sens 26:3515–3526

Jena C, Ghude SD, Pfister GG, Chate DM, Kumar R, Beig G, Surendran DE, Fadnavis S, Lal DM (2015) Influence of springtime biomass burning in South Asia on regional ozone (O_3): a model based case study. Atmos Environ 100:37–47. https://doi.org/10.1016/j.atmosenv.2014.10.027

Junge CE (1962) Global ozone budget and exchange between stratosphere and troposphere. Tellus 14:363–377

Kerckhoffs J, Wang M, Meliefste K, Malmqvist E, Fischer P, Janssen NA, Beelen R, Hoek G (2015) A national fine spatial scale land-use regression model for ozone. Environ Res 140:440–448. https://doi.org/10.1016/j.envres.2015.04.014

Khemani LT, Momin GA, Rao PSP, Vijay Kumar R, Safai PD (1995) Study of Surface Ozone Behaviour at Urban and Forested Sites in India. Atmos Environ 29(16):2021–2024

Khiem M, Ooka R, Huang H, Hayami H, Yoshikado H, Kawamoto Y (2010) Analysis of the relationship between changes in meteorological conditions and the variation in summer ozone levels over the Central Kanto area. Adv Meteorol 2010:13. Article ID 349248

Kim SW, Heckel A, McKeen SA, Frost GJ, Hsie E-Y et al (2006) Satellite-observed U.S. power plant NOx emission reductions and their impact on air quality. Geophys Res Lett 33:L22812. https://doi.org/10.1029/2006GL027749

Kulkarni PS, Ghude SD, Bortoli D (2010) Tropospheric ozone (TOR) trend over three major inland Indian cities: Delhi, Hyderabad and Bangalore. Ann Geophys 28:1879–1885

Kulkarni PS, Bortoli D, Domingues A, Maria Silva A (2016) Surface Ozone Variability and Trend over Urban and Suburban Sites in Portugal. Aerosol Air Qual Res 16:138–152

Kumar P, Imam B (2013) Footprints of air pollution and changing environment on the sustainability of built infrastructure. Sci Total Environ 444:85–101. https://doi.org/10.1016/j.scitotenv.2012.11.056

Kumar R, Naja M, Venkataramani S, Wild O (2010) Variations in surface ozone at Nainital: a high-altitude site in the central Himalayas. J Geophys Res 115(D16):D16302

Lal S, Naja M, Subbaraya BH (2000) Seasonal variations in surface ozone and its precursors over an urban site in India. Atmos Environ 34:2713–2724

Lal S, Venkataramani S, Srivastava S, Gupta S, Mallik C, Naja M, Sarangi T, Acharya YB, Liu X (2013) Transport effects on the vertical distribution of tropospheric ozone over the tropical marine regions surrounding India. J Geophys Res 118. https://doi.org/10.1002/jgrd.50180

Lal S, Venkataramani S, Chandra N, Cooper OR, Brioude J, Naja M (2014) Transport effects on the vertical distribution of tropospheric ozone over western India. J Geophys Res-Atmos 119:10012–10026

Lee H-J, Kim S-W, Brioude J, Cooper OR, Frost GJ et al (2013) Transport of NOx in East Asia identified by satellite and in-situ measurements and Lagrangian particle dispersion model simulations. J Geophys Res (in press)

Lee H-J, Kim S-W, Brioude J, Cooper OR, Frost GJ, Kim C-H, Park RJ, Trainer M, Woo J-H (2014) Transport of NOx in East Asia identified by satellite and in situ measurements and Lagrangian particle dispersion model simulations. J Geophys Res-Atmos 119:2574–2596. https://doi.org/10.1002/2013JD021185

Lee J-B, Cha J-S, Hong S-C, Choi J-Y, Myoung J-S, Park RJ, Woo J-H, Ho C, Han J-S, Song C-K (2015) Projections of summertime ozone concentration over East Asia under multiple IPCC SRES emission scenarios. Atmos Environ 106:335–346

Lefohn AS, Shadwick D, Oltmans SJ (2010) Characterizing changes in surface ozone levels in metropolitan and rural areas in the United States for 1980–2008 and 1994–2008. Atmos Environ 44:5199–5210

Lelieveld J, van Aardenne J, Fischer H, de Reus M, Williams J, Winkler P (2004) Increasing ozone over the Atlantic Ocean. Science 304:1483–1487

Lemaire VEP, Colette A, Menut L (2016) Using statistical models to explore ensemble uncertainty in climate impact studies: the example of air pollution in Europe. Atmos Chem Phys 16:2559–2574

Li J, Yang W, Wang Z, Chen H, Hua B, Li B, Sun Y, Fu P, Zhang Y (2016) Modeling study of surface ozone source-receptor relationships in East Asia. Atmos Res 167:77–88

Lin C, Jacob DJ, Munger W, Fiore A (2000) Increasing background ozone in surface air over the United States. Geophys Res Lett 27:3465–3468. https://doi.org/10.1029/2000GL011762

Lin W, Xu X, Zhang X, Tang J (2008) Contributions of pollutants from North China Plain to surface ozone at the Shangdianzi GAW station. Atmos Chem Phys 8:5889–5898. https://doi.org/10.5194/acp-8-5889-2008

Lin M et al (2012) Transport of Asian ozone pollution into surface air over the western United States in spring. J Geophys Res 117:D00V07. https://doi.org/10.1029/2011JD016961

Lin M, Horowitz LW, Oltmans SJ, Fiore AM, Fan S (2014) Tropospheric ozone trends at Mauna Loa Observatory tied to decadal climate variability. Nat Geosci 7:136–143. https://doi.org/10.1038/ngeo2066

Lin M, Fiore AM, Horowitz LW, Langford AO, Oltmans SJ, Tarasick D, Rieder HE (2015) Climate variability modulates western US ozone air quality in spring via deep stratospheric intrusions. Nat Commun 6:7105

Linvill DE, Hooker WJ, Olson B (1980) Ozone in Michigan's environment 1876–1880. Mon Weather Rev 108:1883–1891

Lisac I, Vujnović V, Marki A (2010) Ozone measurements in Zagreb, Croatia, at the end of 19th century compared to the present data. Meteorol Z 19:169–178

Liu SC, Trainer M, Fehsenfeld FC, Parrish DD, Williams EJ et al (1987) Ozone production in the rural troposphere and the implications for regional and global ozone distributions. J Geophys Res 92(D4):4191–4207. https://doi.org/10.1029/JD092iD04p04191

Logan JA, Prather MJ, Wofsy SC, McElroy MB (1981) Tropospheric chemistry: A global perspective. J Geophys Res 86(C8):7210–7254. https://doi.org/10.1029/JC086iC08p07210

Ma J, Lin WL, Zheng XD, Xu XB, Li Z, Yang LL (2014) Influence of air mass downward transport on the variability of surface ozone at Xianggelila Regional Atmosphere Background Station, southwest China. Atmos Chem Phys 14:5311–5325

Marenco A, Gouget H, Nédélec P, Pagés J-P (1994) Evidence of a long-term increase in tropospheric ozone from Pic du Midi series: consequences: positive radiative forcing. J Geophys Res 99:16,617–16,632

McLinden CA, Olsen SC, Hannegan B, Wild O, Prather MJ, Sundet J (2000) Stratospheric ozone in 3-D models: A simple chemistry and the cross-tropopause flux. J Geophys Res 105:14653–14665. https://doi.org/10.1029/2000JD900124, 2000.

Meehl GA, Stocker TF, Collins WD, Friedlingstein P, Gaye AT, Gregory JM, Kitoh A, Knutti R, Murphy JM, Noda A, Raper SCB, Watterson IG, Weaver AJ, Zhao Z-C (2007) Climate Change 2007: The Physical Science Basis, contribution of working group I to the fourth assessment report of the intergovernmental panel on climate change Global Climate Projections. Cambridge University Press, Cambridge, UK/New York, pp 747–846

Mittal ML, Hess PG, Jain SL, Arya BC, Sharma C (2007) Surface ozone in the Indian region. Atmos Environ 41:6572–6584

Mishra AK, Rai R, Agrawal SB (2013) Individual and interactive effects of elevated carbon dioxide and ozone on tropical wheat (Triticum aestivum L.) cultivars with special emphasis on ROS generation and activation of antioxidant defense system. Indian J Biochem Biophys 50:139–149

Monks PS (2000) A review of the observations and origins of the spring ozone maximum. Atmos Environ 34:3545–3561

Monks PS, Granier C, Fuzzi S, Stohl A, Williams M et al (2009) Atmospheric Composition Change – Global and Regional Air Quality. Atmos Environ 43:5268–5350

Monks PS, Archibald AT, Colette A, Cooper O, Coyle M, Derwent R, Fowler D, Granier C, Law KS, Mills GE, Stevenson DS, Tarasova O, Thouret V, von Schneidemesser E, Sommariva R, Wild O, Williams ML (2015) Tropospheric ozone and its precursors from the urban to the global scale from air quality to short lived climate forcer. Atmos Chem Phys 15:8889–8973

Naik V, Voulgarakis A, Fiore AM, Horowitz LW, Lamarque J-F, Lin M, Prather MJ, Young PJ, Bergmann D, Cameron-Smith PJ, Cionni I, Collins WJ, Dalsøren SB, Doherty R, Eyring V, Faluvegi G, Folberth GA, Josse B, Lee YH, MacKenzie IA, Nagashima T, van Noije TPC, Plummer DA, Righi M, Rumbold ST, Skeie R, Shindell DT, Stevenson DS, Strode S, Sudo K, Szopa S, Zeng G (2013) Preindustrial to present-day changes in tropospheric hydroxyl radical and methane lifetime from the Atmospheric Chemistry and Climate Model Intercomparison Project (ACCMIP). Atmos Chem Phys 13:5277–5298

Nair PR, Chand D, Lal S, Modh KS, Naja M, Parameswaran K, Ravindran S, Venkataramani S (2002) Temporal variations in surface ozone at Thumba (8.6°N, 77°E)-a tropical coastal site in India. Atmos Environ 36(4):603–610

Naja M, Lal S (2002) Surface ozone and precursor gases at Gadanki (13.5 N, 79.2 E) a tropical rural site in India. J Geophys Res 107:4179. https://doi.org/10.1029/2001JD000357

Naja M, Lal S, Chand D (2003) Diurnal and seasonal variabilities in surface ozone at a high altitude site Mt Abu (24.6 N, 72.7 E, 160 ma asl) in India. Atmos Environ 37:4205–4215

Nishanth T, Praseed KM, Kumar SMK, Valsaraj KT (2014) Observational study of surface O_3, NOx, CH_4 and total NMHCs at Kannur, India. Aerosol Air Qual Res 14:1074–1088

Nolle M, Ellul R, Ventura F, Güsten H (2005) A study of historical surface ozone measurements (1884–1900) on the island of Gozo in the central Mediterranean. Atmos Environ 39:5608–5618

Olsen MA, Schoeberl MR, Douglass AR (2004) Stratosphere-troposphere exchange of mass and ozone. J Geophys Res 109:D24114

Oltmans SJ, Johnson BK, Harris JM (2012) Springtime boundary layer ozone depletion at Barrow, Alaska: Meteorological influence, year-to-year variation, and long-term change. J Geophys Res 117:D00R18. https://doi.org/10.1029/2011JD016889

Oltmans SJ, Lefohn AS, Shadwick D, Harris JM, Scheel HE et al (2013) Recent tropospheric ozone changes — A pattern dominated by slow or no growth. Atmos Environ 67:331–351

Pandey J, Agrawal M, Khanam N, Narayan D, Rao DN (1992) Air Pollutant Concentrations in Varanasi, India. Atmos Environ 26B:91–98

Parrish DD, Millet DB, Goldstein AH (2009) Increasing ozone in marine boundary layer inflow at the west coasts of North America and Europe. Atmos Chem Phys 9:1303–1323. https://doi.org/10.5194/acp-9-1303-009.

Parrish DD et al (2010) Impact of transported background ozone inflow on summertime air quality in a California ozone exceedance area. Atmos Chem Phys 10:10,093–10,109. https://doi.org/10.5194/acp-10-10093-2010

Parrish DD, Law KS, Staehelin J, Derwent R, Cooper OR et al (2012) Long-term changes in lower tropospheric baseline ozone concentrations at northern mid-latitudes. Atmos Chem Phys 12:11485–11504. https://doi.org/10.5194/acp-12-11485-2012

Pavelin EG, Johnson CE, Rughooputh S, Toumi R (1999) Evaluation of pre-industrial surface ozone measurements made using Schönbein's method. Atmos Environ 33:919–929

Permadi DA, Oanh NTK (2008) Episodic ozone air quality in Jakarta in relation to meteorological conditions. Atmos Environ 42:6806–6815

Pudasainee D, Sapkota B, Shrestha ML, Kaga A, Kondo A, Inoue Y (2006) Ground level ozone concentrations and its association with NOx and meteorological parameters in Kathmandu valley, Nepal. Atmos Environ 40(40):8081–8087

Pusede SE, Steiner AL, Cohen RC (2015) Temperature and Recent Trends in the Chemistry of Continental Surface Ozone. Chem Rev 115:3898–3918

Reddy BS, Kumar KR, Balakrishnaiah G, Gopal KR, Reddy RR, Sivakumar V, Lingaswamy AP, Arafath SM, Umadevi K, Kumari SP, Ahammed YN, Lal S (2013) Analysis of diurnal and seasonal behavior of surface ozone and its precursors (NOx) at a semi-arid rural site in southern India. Aerosol Air Qual Res 12:1081–1094

Roy SD, Beig G, Ghude SD (2009) Exposure-plant response of ambient ozone over the tropical Indian Region. Atmos Chem Phys 9:5253–5260

Sandroni S, Anfossi D, Viarengo S (1992) Surface ozone levels at the end of the nineteenth century in South America. J Geophys Res 97:2535–2539

Satsangi G, Lakhani A, Kulshrestha PR, Taneja A (2004) Seasonal and diurnal variation of surface ozone and a preliminary analysis of exceedance of its critical levels at a semi-arid site in India. J Atmos Chem 47(3):271–286

Sarkar A, Singh AA, Agrawal SB, Ahmed A, Rai SP (2015) Cultivar specific variations in antioxidative defense system, genome and proteome of two tropical rice cultivars against ambient and elevated ozone. Ecotoxiocol Environ Saf 115:101–111

Sharma P, Jha AB, Dubey RS, Pessarakli M (2012) Reactive oxygen species, oxidative damage, and antioxidative defense mechanism in plants under stressful conditions. Aust J Bot 2012:1–26

Shena L, Mickley LJ (2017) Seasonal prediction of US summertime ozone using statistical analysis of large scale climate patterns. PNAS 114(10):2491–2496

Shindell D, Kuylenstierna JCI, Vignati E, van Dingenen R, Amann M, Klimont Z, Anenberg SC, Muller N, Janssens-Maenhout G, Raes F, Schwartz J, Faluvegi G, Pozzoli L, Kupiainen K, Höglund-Isaksson L, Emberson L, Streets D, Ramanathan V, Hicks K, Oanh NTK, Milly G, Williams M, Demkine V, Fowler D (2012) Simultaneously Mitigating Near-Term Climate Change and Improving Human Health and Food Security. Science 335:183–189. https://doi.org/10.1126/science.1210026

Simpson D, Arneth A, Mills G, Solberg S, Uddling J (2014) Ozone — the persistent menace: interactions with the N cycle and climate change. Curr Opin Environ Sustain 9–10:9–19. https://doi.org/10.1016/j.cosust.2014.07.008

Singh A, Sarin SM, Shanmugam P, Sharma N, Attri AK, Jain VK (1997) Ozone distribution in the urban environment of Delhi during winter months. Atmos Environ 31:3421–3427

Singh AA, Agrawal SB, Shahi JP, Agrawal M (2014) Investigating the response of tropical maize (Zea mays L.) cultivars against elevated levels of O3 at two developmental stages. Ecotoxocology 23:1447–1463

Srivastava S, Lal S, Venkataramani S, Gupta S, Acharya YB (2011) Vertical distribution of ozone in the lower troposphere over the Bay of Bengal and the Arabian Sea during ICARB-2006: effects of continental outflow. J Geophys Res. https://doi.org/10.1029/2010JD015298

Staehelin J, Thudium J, Buehler R, Volz-Thomas A, Graber W (1994) Trends in surface ozone concentrations at Arosa (Switzerland). Atmos Environ 28:75–87

Stevenson DS, Dentener FJ, Schultz MG, Ellingsen K, van Moije TPC et al (2006) Multimodel ensemble simulations of present-day and near-future tropospheric ozone. J Geophys Res 111:D08301. https://doi.org/10.1029/2005JD006338

Sun L, Xue L, Wang T, Gao J, Ding A, Cooper OR, Lin M, Xu P, Wang Z, Wang X, Wen L, Zhu Y, Chen T, Yang L, Wang Y, Chen J, Wang W (2016) Significant increase of summertime ozone at Mount Tai in Central Eastern China. Atmos Chem Phys 16:10637–10650

Tai APK, Mickley LJ, Heald CL, Wu SL (2013) Effect of CO2 inhibition on biogenic isoprene emission: Implications for air quality under 2000 to 2050 changes in climate, vegetation, and land use. Geophys Res Lett 40:3479–3483. https://doi.org/10.1002/grl.50650

Tang Q, Prather MJ, Hsu J (2011) Stratosphere- troposphere exchange ozone flux related to deep convection. Geophys Res Lett 38:L03806

Teixeira E, Fischer G, van Velthuizen H, van Dingenen R, Dentener F, Mills G, Walter C, Ewert F (2011) Limited potential of crop management for mitigating surface ozone impacts on global food supply. Atmos Environ 45:2569–2576. https://doi.org/10.1016/j.atmosenv.2011.02.002

Tiwari S, Rai R, Agrawal M (2008) Annual and seasonal variations in tropospheric ozone concentrations around Varanasi. Int J Remote Sens 9(15):4499–4514

The Royal Society (2008) Ground-level Ozone in the 21st century: Future trends, impacts and policy implications. Royal Society policy document 15/08, RS1276. Available at http://royal-society.org/Report_WF.aspx?pageid57924&terms5groundlevel1ozone

UNEP (2014) Assessment for decision-makers: scientific assessment of ozone depletion: World Meteorological Organization, Global Ozone Research and Monitoring Project—Report No. 56, Geneva.

Varshney CK, Aggarwal M (1992) Ozone pollution in the urban atmosphere of Delhi. Atmos Environ 26:291–294

Varotsos C, Kondratyev KY, Efstathiou M (2001) On the seasonal variations of surface ozone in Athens, Greece. Atmos Environ 35:315–320

Varotsos C, Efstathiou MN, Kondratyev KY (2003) Long term variation in surface ozone and its precursors in Athens, Greece. A forecasting tool. Environ Sci Pollut Res 10:19–23

Verstraeten WW, Neu JL, Williams JE, Bowman KV, Worden JR, Boersma KF (2015) Rapid increases in tropospheric ozone production and export from China. Nat Geosci. https://doi.org/10.1038/NGEO2493

Vingarzan R (2004) A review of ozone background levels and trends. Atmos Environ 38:3431–3442

Volz A, Kley D (1988) Evaluation of the Montsouris series of ozone measurements made in the nineteenth century. Nature 332:240–242

Wang C (2005) ENSO, Atlantic climate variability, and the Walker and Hadley circulations. In: Diaz HF, Bradley RS (eds) The Hadley Circulation: Present, Past, and Future. Kluwer Academic Publishers, Dordrecht, pp 173–202

Wang T, Wei XL, Ding AJ, Poon CN, Lam KS et al (2009) Increasing surface ozone concentrations in the background atmosphere of Southern China, 1994–2007. Atmos Chem Phys 9:6217–6227

Wang Y, Yimg Q, Hu J, Zhang H (2014) Spatial and temporal variations of six criteria air pollutants in 31provincial capital cities in China during 2013–2014. Environ Int 73:413–422

Wilkinson S, Mills G, Illidge R, Davies WJ (2012) How is ozone pollution reducing our food supply? J Exp Bot 63:527–536. https://doi.org/10.1093/jxb/err317

Wu S, Mickley LJ, Jacob DJ, Logan JA, Yantosca RM et al (2007) Why are there large differences between models in global budgets of tropospheric ozone? J Geophys Res 112:D05302. https://doi.org/10.1029/2006JD007801

Xu W, Lin W, Xu X, Tang J, Huang J, Wu H, Zhang X (2016) Long-term trends of surface ozone and its influencing factors at the Mt Waliguan GAW station, China – Part 1: Overall trends and characteristics. Atmos Chem Phys 16:6191–6205

Yang Y, Liao H, Li J (2014) Impacts of the East Asian summer monsoon on interannual variations of summertime surface layer ozone concentrations over China. Atmos Chem Phys 14:6867–6879

Young PJ et al (2013) Pre-industrial to end 21st century projections of tropospheric ozone from the Atmospheric Chemistry and Climate Model Intercomparison Project (ACCMIP). Atmos Chem Phys 13:2063–2090. https://doi.org/10.5194/acp-13-2063-2013

Zhang L et al (2008) Transpacific transport of ozone pollution and the effect of recent Asian emission increases on air quality in North America: An integrated analysis using satellite, aircraft, ozonesonde, and surface observations. Atmos Chem Phys 8:6117–6136

Zheng XD, Shen CD, Wan GJ, Liu KX, Tang J, Xu XB (2011) 10Be/7Be implies the contribution of stratosphere troposphere transport to the winter-spring surface O_3 variation observed on the Tibetan Plateau. Chin Sci Bull 56:84–88

Zhu B, Akimoto H, Wang Z, Sudo K, Tang J, Uno I (2004) Why does surface ozone peak in summertime at Waliguan? Geophys Res Lett 31:L17104

Ziemke JR, Chandra S, Duncan BN, Froidevaux L, Bhartia PK et al (2006) Tropospheric ozone determined from Aura OMI and MLS: Evaluation of measurements and comparison with the Global Modeling Initiative's Chemical Transport Model. J Geophys Res 111:D19303

Ziemke JR, Chandra S, Labow GJ, Bhartia PK, Froidevaux et al (2011) A global climatology of tropospheric and stratospheric ozone derived from Aura OMI and MLS measurements. Atmos Chem Phys 11:9237–9251

Chapter 2
Tropospheric Ozone Budget: Formation, Depletion and Climate Change

Abstract Because tropospheric O_3 is a secondary pollutant, its concentration in the troposphere largely depends upon different variables that play major roles in its in-situ production. Formation of O_3 in the troposphere largely depends upon the emission of it's precursors like nitrogen oxides (NOx) and volatile organic compounds (VOCs) along with a set of favourable meteorological conditions such as high temperature, intense solar radiations, long sunshine hours, wind speed/direction, etc. In addition, tropospheric O_3 concentration also depends upon the stratospheric intrusion which shows significant seasonal and zonal variations. Apart from O_3 formation, the O_3 budget in the troposphere is also determined by O_3 destruction/depletion, which is more prominent in marine boundary layer, characterized by the production of halogen species like chlorine (Cl), bromine (Br) and iodine (I) and their respective oxides. This chapter emphasizes the role of different factors determining O_3 formation/depletion and the conditions which significantly affect the tropospheric O_3 budget. The effect of climate change variables on tropospheric O_3 budget is also discussed.

Keywords Stratospheric intrusion · Halogen · Ozone depletion

Contents

© Springer International Publishing AG 2018
S. Tiwari, M. Agrawal, *Tropospheric Ozone and its Impacts on Crop Plants*,
https://doi.org/10.1007/978-3-319-71873-6_2

1 Introduction

In addition to being an important tropospheric pollutant, O_3 also plays an important role in the earth's atmospheric chemistry owing to its role in initiation of photochemical oxidation processes (Monks et al. 2015). As O_3 is a secondary pollutant, O_3 budget in the troposphere depend either on the downward transport from the stratosphere (Ma et al. 2014) or on in-situ photochemical reactions involving O_3 precursors including NOx and non-methane volatile organic compounds (VOCs), methane (CH_4) or carbon monoxide (Royal Society 2008; Monks et al. 2015). The tropospheric O_3 concentration at a given location depends mainly upon the balance between photochemical production and stratospheric intrusion. Owing to the substantial concentration of O_3 present in the troposphere, it was earlier believed that transport from the stratosphere was the dominant source of O_3 in the troposphere (Danielsen 1968; Fabian and Pruchniewicz 1977; Chatfield and Harrison 1976). However, it has now been established that the major proportion of tropospheric O_3 budget is actually contributed by photochemical oxidation of CO and hydrocarbon catalyses by HOx and NOx (Chameides and Walker 1973; Royal Society 2008). Present day modeling studies have provided sufficiently strong evidences supporting in-situ photochemistry as the dominant source of tropospheric O_3 as compared to stratospheric tropospheric exchange (Wu et al. 2007). Studies have shown that changes in O_3 precursor emissions have the largest effect on O_3 concentration which increased by 23% in the early years of twentyfirst century compared to 1960 (Revell et al. 2015). In the simulation studies performed with the chemistry climate model SOCOL (Solar Climate Ozone Links), a 6% increase in global mean tropospheric O_3 was predicted by the end of twentyfirst century with fixed anthropogenic O_3 precursor emissions of NOx, CO and non- methane VOCs (Revell et al. 2015). The quantity of tropospheric O_3 originating from stratosphere or from in-situ photochemistry may vary in different chemistry climate global models, but agree that photochemical production exceeds the stratospheric flux by factors of 7–15 (Young et al. 2013). The influence of O_3 transported from the stratosphere is considerable at a few background sites where regional and local emissions of O_3 precursors are extremely limited (Logan et al. 2012; Parrish et al. 2012; Oltmans et al. 2013).

Tropospheric O_3 budget also depends upon the meteorological factors which play significant roles in O_3 formation via the precursors (Tiwari et al. 2008; Jacob and Winner 2009). A strong positive correlation between temperature and O_3 concentration has been observed in many polluted regions (Fu and Tai 2015). Temperature not only speeds up the rates of many reactions involved in O_3 photochemistry, but also influences the biogenic emissions of VOCs from vegetation (Jacob and Winner 2009) Temperature also affects the life time of peroxy acetyl nitrate (PAN) and promote its decomposition into NOx, which is an important O_3 precursor (Jacob and Winner 2009). In addition, warmer climate results in an increase in stratospheric O_3 influx, thus affecting O_3 budget (Collins et al. 2003; SPARC-CCM Val 2010). In future, global temperatures are expected to rise which will bring about changes in the emissions of O_3 precursors (IPCC 2013). Therefore a mechanistic understanding of O_3 temperature relationship is important to effectively regulate the O_3 air quality over the coming years (Pusede et al. 2015).

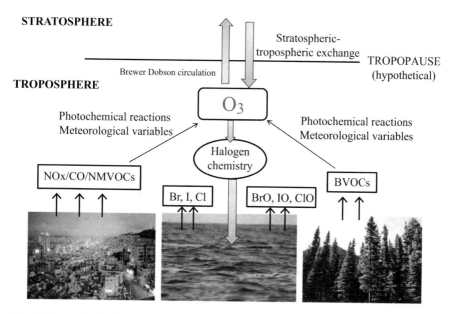

Fig. 2.1 Ozone budget in the troposphere

In addition to O_3 formation, O_3 depletion is also an important factor in determining O_3 budget in the troposphere. Under stable conditions, O_3 depletion occurs either by halogen catalyzed O_3 deposition or NO titration (Monks et al. 2015). A detailed model analysis estimated that halogens are responsible for up to 10% yearly depletion of the total tropospheric O_3 column, especially in the middle and upper troposphere (Saiz-Lopez et al. 2012). Studies have shown that halogens are overall the second most important sink for O_3 in unpolluted and semi-polluted conditions, accounting up to a third of total O_3 loss in the tropical troposphere. The role of halogens in the boundary layer is likely to be impacted by changes in the Earth system, which may alter the formation of marine aerosol, as well as increased emissions of acids and acid precursors, which in turn, affect the halogen activation processes (Long et al. 2014). The tropospheric O_3 budget therefore is determined by a number of factors including in situ chemical production, downward transportation from the stratosphere and O_3 depletion and deposition (Fig. 2.1). The present chapter focuses in details on the various drivers of tropospheric O_3 formation and depletion and the factors that affect the O_3 budget in the troposphere.

2 Ozone Formation in the Troposphere

The important processes that lead to increase in O_3 concentration in the troposphere are discussed in the coming sections.

2.1   Stratospheric – Tropospheric Exchange

Although the main source of O_3 in the troposphere is the photochemical production, the role of stratosphere-troposphere exchange (STE) cannot be overlooked. Several modeling studies done in the last decade have well established the importance of STE in determining the concentration of tropospheric O_3 over a particular region (Škerlak et al. 2014; Ma et al. 2014; Hofmann et al. 2016; Ambrose et al. 2011; Langford et al. 2012; Cooper et al. 2011). It is estimated that global annual O_3 contribution of stratosphere- troposphere exchange may range from 5% to 20% of the total production of tropospheric O_3 (Wild 2007). Hess and Zbinden (2013) have also emphasized the importance of STE in impacting inter-annual O_3 between 30°N and 90°N from 1990 to 2009 using O_3 measurements and global chemical transport model (CAM Chem) for evaluating tropospheric O_3 variability and trends. Lin et al. (2012) showed that it is possible to stimulate temporary O_3 enhancement caused at the surface by the descending stratospheric air masses. Stevenson et al. (2006) studied the future O_3 predictions in 10 chemistry climate models and suggested that the future O_3 budget is greatly influenced by the increase in STE. Most of the climate models predict an increase in the exchange of mass from stratosphere to troposphere under the influence of climate warming however, the amount of increase varies with different models (Stevenson et al. 2006; Shindell et al. 2006; Hegglin and Shepherd 2009). Hegglin and Shepherd (2009) observed that the stratospheric O_3 flux has been increasing at a rate of 2% in Northern Hemisphere since 1970 and by 2100, it is expected to increase the tropospheric O_3 column by 30% compared to 1970. Hess and Zbinden (2013) have observed a high correlation between lower stratospheric and middle tropospheric O_3 which indicates that stratospheric O_3 strongly impacts the trends observed in tropospheric O_3. Observation recorded from Mace Head (Ireland) and Jungfranjoch (Switzerland) measurement sites clearly indicate the important role that STE plays in determining the regional O_3 concentration in the troposphere. At the Mace Head measurement site, baseline O_3 trends have increased between 1987 and 2003 (0.49 ± 0.14 ppbv yr^{-1}) (Hess and Zbinden 2013) although the trend has leveled off in the recent years (Derwent et al. 2007). This unusual trend of tropospheric O_3 at the Mace Head measurement site could not be explained through the emission rates of O_3 precursors, however, the high positive correlation (0.73) between tropospheric and stratospheric O_3 explains 80% of the observed simulated O_3 jump at Mace Head (Hess and Zbinden 2013). Similarly STE can also explain the unusual behavior in tropospheric O_3 trends at Jungfraujoch where a correlation of 0.83 is recorded between lower stratospheric and middle tropospheric O_3 (Hess and Zbinden 2013).

Ambrose et al. (2011) investigated the importance of STE versus long range transport affecting lower tropospheric O_3 concentrations at Mt. Bachelor observatory in central Oregon and reported that STE was the main driver, contributing 52% observed enhanced O_3 level in the region, followed by Asian long range transport whose contribution was 13%. Similar observations were also reported by Cooper et al. (2011) after investigating the contribution to base line O_3 concentration along

California coast over a 6 week period in May/June 2010. It was found that descending stratospheric intrusion had an important influence on O_3 concentration distribution followed by Asian pollution plumes transportation (Cooper et al. 2011). Langford et al. (2012) studied the surface O_3 concentrations at 41 sampling stations and observed that 13% of the variability in maximum 8 h average O_3 in May/June 2010 was attributed to subsidence of O_3 flux from lower stratosphere to upper troposphere.

Lefohn et al. (2011) investigated the occurrence of STE events and their associated O_3 enhancement at 12 O_3 monitoring sites in Northern and Western regions of USA. The work of Lefohn et al. (2011) was extended by Lefohn et al. (2012) who also investigated the frequency of STE events in enhancing hourly O_3 concentrations at 39 urban and rural high and low elevation O_3 monitoring sites in West, Midwest and Eastern US for 2007–2009. The results indicated that STE events contributed to enhance surface O_3 hourly averaged concentrations at the sites across the USA. These enhancements were more frequent at high elevation sites in west and east with preference for springtime (Lefohn et al. 2012).

Hofmann et al. (2016) studied STE events using global and regional model system MECO(n), which couples two different base model: the global model ECHAM 5/MESSy for atmospheric chemistry (Jockel et al. 2010) and the regional atmospheric chemistry and climate model COSMO-CLM/MESSy (Kerkweg and Jockel 2012). It was observed that STE events showed significant zonal and seasonal variations and were maximum during winter and spring (Hofmann et al. 2016). Observations recorded at Waliguan Global Atmospheric Station Watch (GAW) located in North East Tibetan plateau suggested that high O_3 events were mostly observed in spring season when STE frequency was higher (Zheng et al. 2011). The summertime O_3 peak was greatly influenced by the transport of anthropogenic pollutants from central and eastern China (Xue et al. 2011). Similar seasonal variations were also reported by Škerlak et al. (2014) who compiled global climatology of stratospheric O_3 down-flux from 1979 to 2011. Along with the O_3 flux across the tropopause, Škerlak et al. (2014) also studied the phenomenon of 'deep STE', where the stratospheric air reaches the planetary boundary layer within 4 days. While O_3 in the upper troposphere is mainly considered as a 'greenhouse gas' and an oxidizer, deep STE can contribute to the enhanced O_3 levels at the ground (Knowlton et al. 2004). It is therefore important to analyze the transport and mixing of downward stratospheric O_3 flux to clearly evaluate the role of STE in enhancing tropospheric O_3 (Bourqui and Trepanier 2010). The downward O_3 flux across the troposphere peaks in summers when the mass flux shows its minimum value, however, in case of deep STE event, the downward O_3 flux is dominated by mass flux and peaks during the spring time (Škerlak et al. 2014).

In an another study, 2 year measurement of surface O_3 and CO was done from December 2007 to November 2009 at Xianggelila in south west China and it was observed that maximal O_3 and CO mixing ratio were recorded in spring followed by winter and fall, and minimum in summer (Ma et al. 2014). It was observed that Xianggelila was influenced largely by the airflows from southern and northern Tibet plateau in winter, fall and spring which corresponded to high surface O_3 during these seasons. The STE events during winter season are responsible for an increase in winter O_3 by approximately 21% (+9.6 ppb) at Xianggelila (Ma et al. 2014). During summer the movement of O_3 in different layers is greatly influenced by Asian monsoon which creates a condition of continuous cloudiness and frequent precipitation, resulting in minimal O_3 concentration (Ma et al. 2014). Tang et al. (2013) have shown that the Mount Pinatubo eruption in 1991 reduced the flux of O_3 from stratosphere to troposphere until 1995. On the basis of modeling based studies, Olsen et al. (2013) have shown that inter annual variability in O_3 mass flux from the stratosphere to troposphere is approximately 15% in Northern Hemisphere and approximately 6% in Southern Hemisphere. It has been observed that 16% of the Northern Hemisphere mid-latitude tropospheric inter-annual O_3 variability is controlled by the flux from stratosphere (Neu et al. 2014).

In addition to the seasonal variations, STE events are also dominated by Zonal variations. The global hotspots for deep STE are found along the west coast of North America and over the Tibetan Plateau especially during winters and spring (Škerlak et al. 2014). It is important to understand the transport events occurring between lower stratosphere and upper troposphere to evaluate the quantity of O_3 being fluxed into the troposphere from the stratosphere. Moore and Semple (2005) pointed at the existence of Tibetan 'Taylor cap' and a halo of stratospheric O_3 over the Himalayan range, responsible for the elevated levels of upper tropospheric O_3 in the mountainous regions. The high orography combined with high mixing layers that are associated with the Tibetan plateau allow the quasi- horizontal transport of stratospheric O_3 into the planetary boundary layer (Chen et al. 2013). In southern hemisphere, the maximum O_3 flux into the planetary boundary layer was found over the Andes mountain range around 30°S (Škerlak et al. 2014). Certain other events are also known to affect STE events. For instance, Reutter et al. (2015) quantified the importance of extra tropical cyclones and observed that approximately 50–60% of all STE events over North Atlantic occur in the area around the storm/cyclone tracks.

Škerlak et al. (2014) also studied the rapid transport of O_3 into the stratosphere (deep TSE) which is of a common occurrence around the storm track of north Atlantic and north Pacific during all seasons except summers. The STE events dominate in the extratropics regions while TSE events are more predominant in the tropical and polar regions (Škerlak et al. 2014). An analysis of O_3 fluxes over 33 years (1979–2011) showed a positive trend for both deep STE and deep TSE events, however, the downward flux increases twice as fast as the upward flux (Škerlak et al. 2014).

Studies have shown that STE events are also influenced by stratospheric O_3 recovery (Zeng et al. 2010). The STE events are largely influenced by Brewer-Dobson circulation which is characterized by tropospheric air rising into the stratosphere in the tropics, moving pole-wards before descending in middle and high latitudes (Butchart 2014). Stratospheric O_3 recovery accompanied by continuing cooling in the stratosphere and speed up of Brewer-Dobson circulation increased O_3 influx in the troposphere through STE (Shindell et al. 2006; Hegglin and Shepherd 2009). Shindell et al. (2006) reported that stratospheric O_3 recovery may lead to approximately 124% increase in STE in 2100, following in SRES A2 scenario.

Ordonez et al. (2007) also observed a positive trend in tropospheric O_3 over central Europe during 1992–2004 and attributed these changes to the trends of lower stratospheric O_3. UK Chemistry-Aerosol Community (UKCA) model was used to study the stratospheric O_3 changes by the end of the twentyfirst century using WHO A1 scenario of O_3 depleting substances and IPCC A1B scenario for greenhouse gases (Nakicenovic et al. 2000). Based upon this model, stratospheric O_3 is expected to recover between 2040 and 2070 depending upon the latitude. This O_3 recovery is expected to result in large changes in the lower stratospheric O_3 abundance, mostly in the mid latitudes, as a consequence of which large O_3 flux into the troposphere is predicted (Zeng et al. 2010). In the southern hemisphere, where the anthropogenic O_3 production is relatively small, the stratospheric O_3 influx is more important in determining the O_3 budget as compared to northern hemisphere. Zeng et al. (2008) reported a substantial increase of 4–8 ppbv of surface O_3 between 35°S and 90°S during winter months.

Modelling studies done by Zeng et al. (2010) suggested that apart from photochemical production, tropospheric O_3 budget in near future will depend upon stratospheric O_3 recovery and climate change. It has been observed that for the period between 2000 and 2100, the increase in STE amounts to 43% and half of this increase will be due to climate change and other half due to O_3 recovery (Zeng et al. 2010). The effect of increasing STE on troposphere O_3 is counter balanced by increasing water vapours in the future, warm and wetter climate which may promote photochemical destruction of O_3 especially in the tropical lower troposphere. However, owing to the stratospheric O_3 recovery and increased Brewer Dobson circulation, the increase in STE outweighs the increase in photochemical destruction of O_3 resulting in an overall increase in the tropospheric O_3 budget as compared to the present day scenario (Zeng et al. 2010). Banerjee et al. (2016) also emphasized the role of stratospheric ozone recovery and increased Brewer-Dobson circulation in determining the ozone budget. Working on the UKCA model Banerjee et al. (2016) investigated the response of tropospheric O_3 budget to changes in (a) greenhouse gases and climate, (b) O_3 depleting substances (ODSs) and (c) non methane O_3 precursor emissions and observed that climate change and reduction in ODSs will result in increased STE which will offset the observed decreases in net photochemical O_3 production, thus increasing the tropospheric O_3 budget.

2.2 In-situ Photochemistry

In-situ photochemical production of O_3 is the most dominant source of tropospheric O_3 (Monks et al. 2015). Photochemical mechanism for generation of O_3 was first identified by Haagen-Smit (1952). The role of CO, NOx and hydrocarbons as well as the hydroxyl radicals involved in the chemistry of O_3 formation in the troposphere was first proposed by Crutzen (1973) and Chameides and Walker (1973). O_3 photochemistry in the troposphere involved both production and destruction of ozone (Monks 2005; Royal Society 2008). However, the increased anthropogenic activities which result in more and more emissions of O_3 precursors cause more production of O_3 as compared to its destruction (Monks et al. 2015). O_3 production in the atmosphere occurs by the following processes:

(i) Oxidation of CO by hydroxyl radicals (OH°).
(ii) Oxidation of methane and non-methane hydrocarbons in the presence of nitrogen oxides (NO_x).
(iii) Photolysis of NO_2 and the subsequent association of the photoproduct with atmospheric O_2.

OH° radicals play an important role in tropospheric O_3 chemistry (Royal Society 2008) OH° radicals are the important oxidative species responsible for the oxidative potential of the atmosphere (Gligorovski et al. 2015). These radicals act as the initiators of chain reactions in the atmosphere, especially OH° polluted troposphere which controls the mixing ratio and vertical profile of OH° radicals (Stone et al. 2012). The OH° concentrations are strongly affected by O_3 and in turn they substantially affect O_3 concentration (Logan et al. 1981). In the upper troposphere, OH° radicals are produced by the photolysis of gaseous O_3 at UV wavelengths shorter than 320 nm as given in reaction R1 & R2 (Rohrer and Berresheins 2006)

$$O_3 + hv \rightarrow \left[O^*\right] + O_2 \tag{2.1}$$

$$\left[O^*\right] + H_2O \rightarrow OH° + OH° \tag{2.2}$$

[O*]=Electronically excited state atomic oxygen

In the lower troposphere, the formation of OH° radicals by photolysis of O_3 is not much prominent due to weaker overlapping between O_3 absorption band and solar spectrum (Rohrer and Berresheins 2006). Li et al. (2008) reported the formation of nitrous acid (HONO) in the lower troposphere through the reaction between electronically excited $NO_2(NO_2^*)$ and water vapours via the following reactions:

$$NO_2 + hv \rightarrow NO_2^* \tag{2.3}$$

$$NO_2^* + H_2O \rightarrow OH° + HONO \tag{2.4}$$

However, Carr et al. (2009) observed that this conversion of NO_2 to HONO is feasible only in the boundary layer of polar regions especially when the sun is lying low on the horizon. In addition to this, a small proportion of OH° induced O_3 formation is also initiated by the reaction of O_3 with more complex VOCs like alkanes (Zhou et al. 2014).

$$C_2H_6 + OH + O_2 \rightarrow C_2H_5O_2 + H_2O \tag{2.5}$$

$$C_2H_5O_2 + NO \rightarrow C_2H_5O + NO_2 \tag{2.6}$$

Kim et al. (2014) suggested that HONO are important source of OH radicals in the troposphere

$$HONO + hv \rightarrow NO + OH^{\circ} \tag{2.7}$$

Some studies have suggested that the contribution of HONO photolysis to integrated OH radical yield is 60% (Kleffmann et al. 2005; Aumont et al. 2003). Sorgel et al. (2011) demonstrated that even in unpolluted forested regions the contribution of HONO photolysis to OH° generation is higher that the contribution of O_3 photolysis (2.1 and 2.2). Further, it was observed that in highly polluted cities such as Santago de Chile, the most primary source of OH° radicals is the photolysis of HONO (81 and 52% during summer and winter, respectively), followed by ozonolysis of alkanes (2.5 and 2.6) (12 and 29% in winter and summer, respectively), photolysis of formaldehyde (HCHO) (6.1 and 15% in winter and summer, respectively) and then photolysis of O_3 (<1 and 4% in winter and summer, respectively) (Elshorbany et al. 2010).

The conversion efficiency of [O*] to OH° depends on the relative rates of the reactions Eqs. 2.1 and 2.2. Water vapours are the important tropospheric entities in determining the rates of Eqs. 2.1 and 2.2 and therefore its concentration in troposphere is important for OH° radical formation (Royal Society 2008). Water vapour content of the atmosphere is largely dependent on the temperature and relative humidity components of climate change. For instance, in water vapour saturated atmosphere, the fraction of [O] involved in Eq. 2.2 increases from 9% at 10°C C to about 12% at 15°C (Gligorovski et al. 2015). The OH° radicals formed react with other tropospheric pollutants like carbon monoxide (CO), methane (CH_4) and non methane hydrocarbons (NMHC) resulting in their oxidation and subsequently O_3 formation. The oxides of nitrogen (NO_x) play an important role in OH° radical induced oxidation of CO, CH_4 and NMHC (Fig. 2.2). In addition, the photolysis of NOx (NO_2) can also lead to the formation of O_3 through the following reactions:

$$NO_2 + hv \rightarrow NO + [O] \tag{2.8}$$

$$[O] + O_2 \rightarrow O_3 \tag{2.9}$$

The oxides of nitrogen which acts as one of the important precursors of tropospheric O_3 give a complicated outlook to O_3 chemistry due to the manifold ways in which NO and NO_2 are inter-converted.

$$HO_2^{\circ} + NO \rightarrow NO_2 + OH^{\circ} \tag{2.10}$$

$$O_3 + NO \rightarrow NO_2 + O_2 \tag{2.11}$$

Considering reactions Eqs. 2.8, 2.9 and 2.11 a ratio of (NO)/(NO_2) can be estimated. It has been observed that this ratio depends upon the local concentration of O_3, the rate coefficient of Eq. 2.9 and photolysis frequency of Eq. 2.8. This ratio is

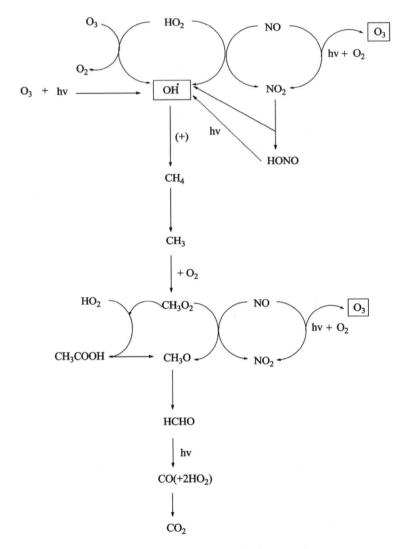

Fig. 2.2 In-situ photochemistry leading to ozone formation in troposphere

termed as Leighton ratio and facilitates the derivation of an expression for the equilibrium concentration of O_3 (Leighton 1961). Although these reactions are important sources of O_3 formation in the troposphere, they do not represent the mechanism for the net production of O_3 in the troposphere (Monks et al. 2015). Other O_3 precursors like VOCs (CH_4 and CO) are also oxidized by OH° radicals leading to the formation of hydroxyl peroxy radicals (HO_2°) and CH_3O_2, further reacts with NO and lead to the inter-conversion of NO to NO_2 via the following reactions:

$$OH^{°} + CO \rightarrow H + CO_2 \tag{2.12}$$

$$H + O_2 \rightarrow HO_2^{°} \tag{2.13}$$

$$OH^{°} + CH_4 \rightarrow CH_3^{°} + H_2O \tag{2.14}$$

$$CH_3^{°} + O_2 \rightarrow CH_3O_2^{°} \tag{2.15}$$

$$CH_3O_2^{°} + NO \rightarrow CH_3O^{°} + NO_2 \tag{2.16}$$

$$HO_2^{°} + NO \rightarrow OH^{°} + NO_2 \tag{2.17}$$

$$NO_2 + hv \rightarrow NO + [O] \tag{2.18}$$

The presence of VOCs and NOx allow $OH^{°}$ to be regenerated and promote the formation of O_3 through photolysis of NO_2. The relationship between VOCs and NOx and their respective roles in O_3 formation is discussed in Sect. 2.2.

The hydroxyl peroxy radicals can further react with O_3 forming $OH^{°}$ radicals

$$HO_2^{°} + O_3 \rightarrow OH^{°} + 2O_2 \tag{2.19}$$

$$OH^{°} + O_3 \rightarrow HO_2^{°} + O_2 \tag{2.20}$$

$HO_2^{°}$ radicals can also recombine together forming H_2O_2, which in turn acts as a source of $OH^{°}$ radicals by the photolytic cleavage of the O-H bond

$$HO_2^{°} + HO_2^{°} \rightarrow H_2O_2 + O_2 \tag{2.21}$$

$$H_2O_2 + OH^{°} \rightarrow HO_2^{°} + H_2O \tag{2.22}$$

$HO_2^{°}$ formed can also react with NO leading to the catalytic production of NO_2.

$$HO_2^{°} + NO \rightarrow OH^{°} + NO_2 \tag{2.23}$$

NO_2 thus formed acts a source of O_3 via the reactions Eqs. 2.8 and 2.9 as these reactions suggest $OH^{°}$ to play an important role in tropospheric O_3 chemistry. In addition, concentration of NOx also plays an important role in determining the course of O_3 production. In urban areas, where the concentrations of NOx and VOCs are high; both the precursors play important roles in O_3 formation (Fig. 2.3). However, in non urban areas, where the concentration of NOx is comparatively lower, NOx acquires the role of the predominant O_3 precursor responsible for the formation of O_3 in the troposphere (Royal Society 2008) (Fig. 2.3).

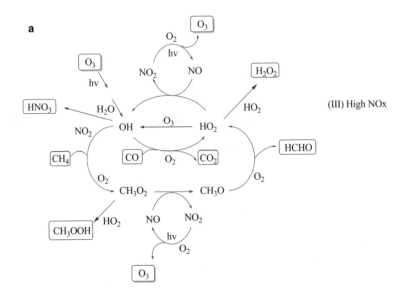

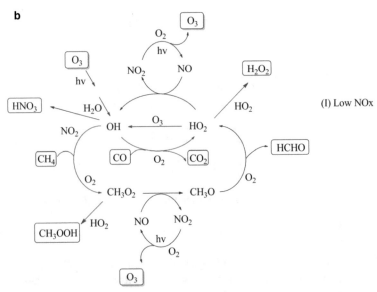

Fig. 2.3 (**a**) Role of high NOx in ozone formation in troposphere (**b**) Role of low NOx in ozone formation in troposphere

2.3 Role of O₃ Precursors

The concentration of O_3 precursors (viz NOx, VOCs, CO etc.) is an important factor in determining the O_3 concentration in that region. The spatial and temporal distribution of O_3 and its precursor are mostly driven by the distribution of their emissions. It is essential to gain accurate knowledge of the surface emission and their evolution with time to understand the analysis and modeling of our quality and climate change interactions. Several global model studies have investigated the role of O_3 percussions and how changes in these drivers could affect the troposphere O_3 budget, from pre industrial era to future projections (Lamarque et al. 2011; Kawase et al. 2011). Future projections of anthropogenic O_3 precursor emissions adopted high populations/high fossil fuel growth scenarios of Nakicenovic et al. (2000) which predicts large increase in troposphere O_3 levels (Zeng and Pyle 2003); Shendell et al. 2006). However, a few other projections have included scenarios with more extensive air quality legislation resulting in reductions in anthropogenic precursor emissions and subsequent decreased tropospheric O_3 as compared to the present day (West et al. 2013; Kawase et al. 2011; Lamarque et al. 2011). As far as the natural emission sources of O_3 precursors are concerned, it has been observed that lightening acts as a major source of NOx emissions which is generally thought to increase in warmer climate (Schumann and Huntrieser 2007), however this prediction is not universal (Jacobson and Streets 2009). In addition to this, bacterial and volcanic activities and soil NOx emissions are also important natural sources contributing to the atmospheric NOx budget. Global estimates of natural NOx emission are highly uncertain ranging from 10 to 60 Tg Ny^{-1} depending upon the sources which are considered (Royal Society 2008). Anthropogenic sources of NOx, include fossil fuel combustion in electric utilities, high temperature operations in several industrial sources and operation of motor vehicles and contributes approximately 33 Tg Ny^{-1} globally (Cofala et al. 2007). NOx not only plays an important role in direct formation of O_3, it also reacts with hydrocarbons, adding up the tropospheric O_3 budget (Monks et al. 2015). In addition to this, NOx also produces nitrous acid which acts as an important source of $OH°$ radicals, influencing the tropospheric O_3 chemistry (Kim et al. 2014). Due to the multiple role of NOx in O_3 chemistry, it is expected that the predicted variations in O_3 concentrations largely depended on the variations in NOx (Tagaris et al. 2007). As far as the consequences of global climate change are concerned, NOx emissions are not much influenced since they are largely anthropogenic in nature and do not shows any strong function of temperature (Tagaris et al. 2007). Owing to the EPA's clean Air Interstate Rule (CAIR) which requires the power plants to reduce their NOx emissions by 60% during the next two decades, as compared to 2003 emissions, and the implementation of new standards for off road diesel equipments, it has been predicted that NOx emissions will decline in the near future (Ibrahim 2014). Tagaris et al. (2007) estimated a total decline in the NOx emissions to be 14% between 2001 and 2020. It has been recorded that a reduction of 18% in the NOx emissions between 2000 and 2006 has already been recorded (EPA 2007). However, as the land based emissions of NOx are observed to decline, shipping sources of NOx may become significant (Eyring et al. 2010; Dalsoren et al. 2010). Dalsoren et al. (2010) showed that 0.5–2.5 ppbv increase in the yearly average surface O_3 concentration is attributed to the emissions of NOx from shipping sources. Transport emissions are predicted to be significant contributor to US and European O_3 budget by 2050 (Hauglustaine and Koffi 2012). Monks et al. (2015) have observed very large increase in NOx emissions in China over the past two years. This increase can be attributed to the increased use of coal for energy generation and industrial activities as well as to a large increase in the number of vehicles (Kurokawa et al. 2013). Studies have shown that the implemented control measures in China has lead to decrease in the emissions of most of the pollutants, but NOx emission is still a problem (Zhao et al. 2013).

NOx affects the O_3 budget by yet another phenomenon termed as "NOx titration" which states that local coupling of NOx and O_3 is important as the reduction in NO can contribute to increase in O_3 (Sicard et al. 2013). The spatial variability in O_3 in large urban area to is generally driven by the phenomenon of "NOx titration" (Escudero et al. 2014). Owing to this strong NOx-O_3 relationship, the NO_2: NO emission ratio caused by the increase in the share of diesel vehicles play an important role in determining the formation of O_3 in a particular area (Weiss et al. 2012; Carslaw et al. 2011).

Volatile organic compounds (VOCs) are yet another class of O_3 precursors that participate actively in tropospheric O_3 chemistry. VOCs can be of both anthropogenic and natural biogenic origin. Studies with global transport model states that the contribution of biogenic VOCs to O_3 generation amounts to about 40–70% of the total contribution of all chemical O_3 precursors in the troposphere (Xie et al. 2008). Isoprene is the most important contributor of biogenic VOC emission (BVOCs) and accounts to emission of approximately 500–600 Tgy^{-1} of BVOCs (Royal Society 2008). Methane is yet another important VOC and has different natural emission sources. Wetlands, oceans and termites are important natural emission sources of methane accounting for approximately 100–230 Tgy^{-1}, 10 Tgy^{-1} and 20 Tgy^{-1} of methane emissions (Royal Society 2008). Out of the anthropogenic VOC emission, aromatic hydrocarbons make the greatest contribution to O_3 formation (Berezina et al. 2017). Accounting to the current estimates, the share of aromatic hydrocarbons contributing towards O_3 generation by oxidation of anthropogenic VOCs is approximately 40% of total O_3 produced (Dreyfus et al. 2002). Based upon this estimations, it is important to estimate the increase in the emission of aromatic hydrocarbons by transport vehicles, fuel and energy enterprises and chemical and metallurgical industries. This estimation is important to analyze the contribution of aromatic hydrocarbons to photochemical O_3 generation and to forecast extreme environmental situations (Berezina et al. 2017). GAW WMO network of stations (Global Atmospheric watch of the world meteorological organization) is an important network involved in the monitoring of VOCs, which include isoprene's and aromatic hydrocarbons such as benzene, toluene, methylbenzene and xylenes (GAW report 2012). In Russia, benzene and toluene concentrations are sporadically measured at the stations of state Air Pollution control network (Bezuglaya & Smirnova 2008). In yet another study in Russia, the contribution of aromatic VOCs to O_3 formation was analyzed in different regions along the Trans-Siberian Railways from Moscow to Vladivostok using the TRO-ICA-12 Mobile laboratory (Berezina et al. 2017). The results of this study indicated a considerable contribution (~30–50%) of anthropogenic emissions of VOCs to photochemical O_3 generation in large cities along the Trans-Siberian Railway (Berezina et al. 2017).

Terrestrial vegetation acts as a dominant source of emission of atmospheric VOCs (Guenther et al. 1995). Along with the other compounds, isoprenes and monoterpenes are the most important VOCs emitted. MEGAN, a model system calculating temporal and spatial rates of emissions of chemical compounds from terrestrial ecosystems to the atmosphere under varying environmental conditions, has been used in several studies to quantify the emission of biogenic VOCs to the atmosphere, the most recent version of model being MEGAN v2.1 (Guenther et al. 2012). Based upon this model, it has been estimated that isoprene emissions account for 56% of the total VOC emissions (Guenther et al. 2012). Previous studies have shown significant differences between the total amounts of BVOCs emitted by vegetation (Sindelarova et al. 2014). These differences can be attributed to different factors affecting BVOC emissions such as radiations amount, leaf temperature, sol moisture, vegetation type and the distribution of plant functional types used in different models (Monks et al. 2015). Studies conducted using the LPJ-GUESS model have shown that variations in meteorology and vegetation description among different models may lead to substantially different BVOCs emission estimates (Arneth et al. 2011).

Chatani et al. (2014) studied the emission of NOx and VOCs over east and south Asia and compared the data obtained in 2010 (through surface monitoring, ozone-sondes and satellites) with those predicted for 2030 [through the regional air quality simulation framework including the weather Research and forecasting Modeling system (WRF) and the community Multi-scale air quality modeling system (CMAQ)]. VOC emission rate was found to increase by 27.1% and 94.1% respectively, in China and India, whereas NOx emission increased by 35.6% in China and 24.8% in India between 2010 and 2030. (Chatani et al. 2014). An analysis of VOC/NOx emission ratio suggests that the increasing tropospheric O_3 budget in India is largely influenced by NOx emissions while those of China is governed mainly by the VOC emissions (Chatani et al. 2014). O_3 chemistry largely depends upon the emission ratio of NOx and VOCs (Royal Society 2008). Under the conditions of low NOx levels, O_3 formation is referred to as 'NOx limited' or 'NOx-sensitive' because O_3 formation rate increases with increasing NOx concentration (Fig. 2.3). Low NOx conditions promote the formation of OH° radicals which are an important component of O_3 formation reactions. In this condition O_3 formation rate is insensitive to changes in the concentration of VOC. Contrary to this under high NOx emission levels, increasing NOx levels serve to inhibit O_3 formation by scavenging the OH° radicals. In this case the role of VOCs in O_3 formation becomes more prominent (Fig. 2.3b) (Royal Society 2008). This condition is also referred to as 'VOC-limited' or 'VOC-sensitive'. VOCs, specifically isoprene acts as a precursor for O_3 in polluted, high- NOx regions, but shows a tendency to reduce O_3 by ozonolysis in remote low- NOx regions (Fu and Tai 2015).

Isaksen et al. (2014) studied the effect of permafrost thawing on the VOC, especially methane emissions in the Arctic regions. As the increase in temperature is one of the most important outcome of climate change (IPCC 2013), permafrost thawing in the arctic regions becomes an unavoidable feature. Methane is likely to be released from the Arctic as a result of thawing of permafrost in near future (Shakhova et al. 2009). Shakhova et al. (2010) stated that the yedoma region of Siberia contains about 2–5% of organic carbon in the upper 25 m of the permafrost which acts as a source of methane. In addition to this, methane could be released from methane hydrates on the Arctic shelf during thawing in warmer future climate (Shakhova et al. 2010). Estimates for the permafrost organic carbon in yedoma region of Siberia suggest that this pool contains twice the amount of current atmospheric carbon (Strauss et al. 2013). 30% of the deposited organic carbon is expected to be converted to methane and released into the atmosphere after thawing (Schuur et al. 2015). Using Oslo chemistry transport model (Oslo CTM2) the amount of methane required to be released by the Arctic permafrost thawing to raise the atmospheric methane was calculated (Isaksen et al. 2014).

It was observed that increase in the global emission of methane by 1.9, 3 and 4.8 Gty^{-1} through permafrost thawing will lead to an increase of 3, 6 and 12% of the atmospheric methane budget. Although the required methane emissions are significantly higher than the current methane emissions ($\sim$0.5 Gty^{-1}), under current climate change scenario, it is predicted that thawing of Arctic permafrost could serve as a significant source of the deficient methane emissions (Isaksen et al. 2014). The calculations further suggest that an increase in the atmospheric methane by 3, 6 and 12% will result in an increase in the tropospheric O_3 by 50, 70 and 120 ppb, respectively (Isaksen et al. 2014). In addition to methane, terpenoids (a large group of biogenic VOCs) also contribute a major share in O_3 formations in Arctic regions. (Lindwall et al. 2015). Terpenoids can either be derived directly from *de-novo* synthesis or can be released from the storage pool (Taipale et al. 2011). The emission of *de-novo* synthesized compounds is light and temperature dependant while that released from storage pools is mainly dependant on leaf temperature (Ehn et al. 2014). Therefore, the emission of BVOCs often peaks during midday and decrease with decreasing light and temperature (Mckinney et al. 2011). However in the Arctic region long sunshine hours during midnight period suggest that significant amounts of biogenic VOCs may be released from the Arctic ecosystem. Lindwall et al. (2015) compared BVOC emissions the high of arctic, low arctic and sub arctic vegetation and observed that low arctic and sub arctic sites released more BVOCs during 24 h. period with night time emissions in the same range as those during the day. This study by Lindwall et al. (2015) clearly suggests that the night emissions of BVOCs in the arctic ecosystems cannot be ignored.

2.4 Effect of Meteorological Variables

Due to the photochemical nature of O_3 production in the troposphere, it is strongly influenced by meteorological variables like solar radiation fluxes, temperature, cloudiness, wind speed/directions etc. (Otero et al. 2016; Jacob and Winner 2009). Several studies have indicated the important roles of climate variables on future air

Table 2.1 Effect of different climatic variables on O_3 formation processes

	Climatic variables	Impacts on O_3 formation processes
1.	Temperature	Influences Chemical reaction
		Alters photolytic rates
		Modulates biogenic emissions
		Influences the volatility of chemical species
		Affects Brewer Dobson Circulation (brings about changes in stratospheric tropospheric exchange of O_3)
		Increases the frequency of lightening
2.	Humidity	Modulates $OH°$ radicals
		Precipitations leads to wash down of O_3 precursors
		Affect O_3 chemistry, particularly during spring season
3.	Wind speed and direction	Determines horizontal transport and vertical mixing of chemical species
		Influences dust and sea-salt emissions
4.	Lightning	Determines free troposphere NOx emissions
		Alters the oxidizing capacity of the troposphere
5.	Increasing CO_2 concentration	Influences stratospheric tropospheric exchange of O_3
		Modulates the photochemical reaction of NOx in upper troposphere.

quality in addition to changing anthropogenic emissions (Fiore et al. 2012; Wu et al. 2012; Tai et al. 2013). A better understanding of how the different meteorological variables have interacted in past would be particularly important in analyzing the course of atmospheric chemical evolution in coming decades. For instance, atmospheric circulation controls the short and long term transport of O_3 (Demuzere et al. 2009). In addition to this, the air circulation can also affect the interaction among O_3 precursors, interfering in the process of O_3 formation and destruction (Saavedra et al. 2012). The relationship between surface O_3 and meteorological variables is complex and non-linear (Pusede et al. 2015), but it is usually strongest during the summer season accompanied by high temperatures, peak solar radiations and stagnant conditions (Coates et al. 2016; Anderson and Engardt 2010; Jacob and Winner 2009). The high O_3 levels during the European heat waves of 2003 can be explained by the aforementioned climate conditions (Coates et al. 2016; Otero et al. 2016; Solberg et al. 2008). The important meteorological factors affecting the troposphere O_3chemistry are given in Table 2.1.

Otero et al. (2016) studied the relationship between local and synoptic meteorological conditions and surface O_3 concentration over Europe in spring and summer months during the period 1998–2012 using a novel implementation of traditional objective of Jenkinson and Collinson (1977). This study clearly indicated that meteorological drivers accounted for most of the explained variance of O_3 over central and north-west Europe. It was observed that daily maximum temperature was one of the main drives of O_3 and was positively correlated to it during the summer months over central and north-west Europe (Otero et al. 2016). The high sensitively of O_3 in this region towards temperature can be attributed to the effect of temperature on VOC emissions (Otero et al. 2016). Temperature plays an important determining factor in O_3 formation in southern and northern Europe also, however with a lesser magnitude (Otero et al. 2016). Relative humidity and solar radiations are other important drives of O_3 formation, their role being more prominent during the spring season. This study also emphasized the importance of wind speed and direction in influencing surface O_3 concentration however; their effect was confined to restricted locations in Europe (Otero et al. 2016). As temperature is the most potential meteorological drivers of O_3 formation, a mechanistic understanding of O_3 –temperature relationship is essential to effectively regulate the O_3 air quality over the coming decades. Temperature can influence O_3 production in the troposphere in different ways:

(i)   By altering the troposphere O_3 chemistry especially $OH°$ radical production and concentration (Pusede et al. 2015).
(ii) Increased biogenic VOC emissions (Pacifico et al. 2012).
(iii) Reduced life time of peroxyacetyl nitrate (PAN) due to accelerated decomposition of PAN into NOx at higher temperature (Jacob and Winner 2009).
(iv) Increasing the O_3 mixing ratio (Steiner et al. 2006).

The impact of temperature of O_3 chemistry in the troposphere is largely dependent on the fact that whether the chemistry is NOx limited or VOC limited (discussed in the earlier section of the chapter). If the chemistry is VOC limited, then the increase in temperature stimulates the reactions of O_3 formations, however, if the chemistry is NOx limited, temperature variations have lesser effect (Pusede et al. 2015). In remote areas, where methane dominates the concentration of O_3 precursors, temperature dependence of the oxidation reaction of methane by $OH°$ assumes more importance. As discussed earlier, production and fate of $OH°$ radicals depicts well marked variation under VOC limited and NOx limited conditions (Royal Society 2008). Under conditions that are VOC limited, increasing $OH°$ radicals enhances the O_3 chemistry linearly, whereas under NOx limited conditions, O_3 chemistry varies as square root of $OH°$ radical production (Pusede et al. 2015).

Temperature significantly affects the biogenic emission of VOCs especially isoprene (C_5H_8) which acts as an important source of organic reactivity on warm and hot days in many remote, rural and even urban locations (Ormeno et al. 2010). Enzymatic isoprene synthesis is temperature dependant, increasing with the increased ambient temperature (Pacifico et al. 2012). However at temperature > ~40 °C, isoprene synthase denatures, slowing down the otherwise exponential increase (Alves et al. 2014; Pesude et al. 2015). In addition to isoprene, monoterpenes ($C_{10}H_{16}$) and sesquiterpenes ($C_{15}H_{24}$), emitted by conifers and flowering plants also show exponential increase in response to increasing temperature (Schollert et al. 2014; Mc Kinney et al. 2011; Loathawomkitkul et al. 2009). Increase in the emissions of biogenic VOCs is significant from O_3 formation perspective as it not only increases the organic reactively of O_3 chemistry in the troposphere but also prolongs the lifetime of O_3 precursors like methane (Pacifico et al. 2012). In addition to this, the biogenic VOCs also affect the cloud formation processes which may indirectly affect the intensity of solar radiations which is an important meteorological derives of O_3 formations (Ehn et al. 2014). Coates et al. (2016) examined the non-linear relationship between O_3, NOx and temperature using idealized box model with different chemical mechanism models viz. MCMv3.2 (Master Chemical Mechanisms), CRIv2 (Common Representative Intermediates), MOZRAT-4 (Model for Ozone and Regional Chemical Tracers), RADM2 (Regional Acid Depositions Model) and CBO5 (Chemical Bond Mechanism). The output of these chemical mechanisms suggested that the highest mixing rations of peak O_3 were produced at high temperatures and moderate emissions of NOx regardless of the temperature dependence of isoprene emissions. On the other hand, the least amount of peak O_3 was recorded with low emissions of NOx over the whole temperature range ($15 - 40°C$) when using both temperature independent and temperature dependant source of isoprene emissions (Coates et al. 2016). This study by Coates et al. (2016) clearly indicated that strong reductions in NOx emissions are necessary to counter-act the increase in O_3 pollution at higher temperatures in near future, especially in the urban areas containing a significant amount of isoprene emitting vegetation. Pusede et al. (2014) also established a non-linear relationship of O_3 mixing ratio with NOx and temperature using an analytical model over San Joaquin valley, California. Under high NOx conditions, O_3 concentration registered an increase by 20 ppbv between the temperature range of 20–40°C due to temperature induced faster reaction rates while increased isoprene emissions added up to a further 11 ppbv of O_3 (Coates et al. 2016) Modeling studies predicted a 2 ppbv increase in O_3 per degree Celsius over central Europe (Coates et al. 2016).

3 Ozone Destruction

O_3 destruction in the troposphere occurs through deposition or through the photo-chemical processes, most important being O_3 photolysis. It has been estimated that a large proportion of stratospheric O_3 loss occurs in the tropical marine boundary layer (Reed et al. 2017) and is believed to be a result of high O_3 photolysis rates in the presence of high temperature and water vapour content (Read et al. 2008). O_3 is predominantly formed in continental regions where there are sufficient sources of NOx and is mostly lost in marine regions where NOx sources are limited (Read et al. 2008). In addition to this, a substantial amount of O_3 is lost by its catalytic destruction via the atmospheric halogen specifically bromine and iodine (Shrewen et al. 2017; Parrella et al. 2012). In the past few decades, high reactivity of atomic halogen radicals (viz. Cl, Br, I) and halogen oxides (viz ClO, BrO, IO)) has been established by different workers and their role in influencing the chemical composi-tion of troposphere has been studied (Simpson et al. 2015) (Fig. 2.4).

Tropospheric halogen species play an important role in the chemistry and oxidiz-ing capacity of the troposphere (Wang et al. 2015; Saiz-Lopez and Glasow 2012). Reactive halogens cause depletion of troposphere O_3 though catalytic cycles, affect the partitioning of HOx and NOx, oxidize Dimethyl Sulphide (DMS) and affect the oxidation of volatile organic compounds (Abbatt et al. 2012). Halogen chemistry affects the O_3 lifetime reducing from 26 to 22 days (Sherven et al. 2016). GEOS-

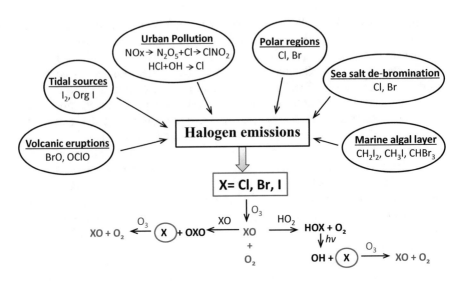

Fig. 2.4 Sources of halogen emissions and halogen chemistry leading to O_3 depletion

Chem chemical transport model observed significant halogen driven changes in concentration of oxidants such as tropospheric O_3 and $OH°$ radicals which decreased by 18.6 and 8.2% respectively (Sherwen et al. 2016). In the same study the effect of halogen chemistry on the lifetime of methane was also estimated. The troposphere lifetime of methane due to $OH°$ (without halogen) was calculated to be 7.47 years. With the inclusion of halogen chemistry, the effect on $OH°$ concentration drop lead to an increase of 10.8% in the lifetime of methane (now 8.28 years). However, with the inclusion of halogen chemistry, the chlorine radicals induced oxidation of methane reduced its lifetime by 1.52%, shortening the lifetime to 8.16 years (Sherwen et al. 2016). Long et al. (2014) have also demonstrated that Cl and Br species significantly affect the temporal and spatial sensitivity of primary atmospheric antioxidants like O_3, HOx, methane, non-methane hydrocarbons etc. In another study, halogen (Br-Cl) radical chemistry as investigated through GEOS- Chem global transport chemistry model results in 14 and 11% decrease in global burdens of tropospheric O_3 and $OH°$ radicals, respectively and a 16% increase in the atmospheric lifetime of methane (Schmidt et al. 2016). The variations observed in the predictions of Sherwen et al. (2016) and Schmidt et al. (2016) can be explained by the fact that the later study does not include sea salt de-bromination phenomenon.

The depletion of O_3 is catalyzed by halogen through the following reactions

$$X + O_3 \rightarrow XO + O_2 \qquad (2.24)$$

$$Y + O_3 \rightarrow YO + O_2 \qquad (2.25)$$

$$XO + YO \rightarrow X + Y + O_2 \qquad (2.26)$$

Net: $2O_3 \rightarrow 3O_2$ Where X and Y represent halogens such as Cl, Br or I.

Equation 2.24 can be involved in other reactions as well:

$$X + O_3 \rightarrow OXO + Y \tag{2.27}$$

$$\rightarrow XY + O_2 \tag{2.28}$$

However photolysis of halogen dioxide formed in Eq. 2.24 can decrease the efficiency of O_3 depletion.

$$OXO + hv \rightarrow XO + [O] \tag{2.29}$$

If X = I, Equation 2.27 will have different products.

$$OIO + hv \rightarrow I + O_2 \tag{2.30}$$

OXO can also react with NO. The rate of co-efficient of this reaction increases as I > Br > Cl.

$$OXO + NO \rightarrow XO + NO_2 \tag{2.31}$$

Equation 2.26 dominates O_3 destruction when BrO and IO mixing ratios are sufficiently high. At lower mixing ratios the reaction with HO_2 radicals predominates:

$$XO + HO_2 \rightarrow HOX + O_2 \tag{2.32}$$

$$HOX + hv \rightarrow X + OH^0 \tag{2.33}$$

$$OH^0 + CO[+O_2] \rightarrow HO_2 + CO_2 \tag{2.34}$$

$$X + O_3 \rightarrow XO + O_2 \tag{2.35}$$

$$CO + O_3 \rightarrow CO_2 + O_2 \tag{Net}$$

Read et al. (2008) gave substantial evidence to prove that halogens have a significant impact on regional as well as global O_3 levels. Spectroscopic measurements at Cape Verde observatory (located within the tropical Eastern North Atlantic Ocean) indicated that, mean daily observed O_3 loss is ~50% greater than that stimulated by a global chemistry model using a classical photochemistry scheme that excluded halogen chemistry (Reed et al. 2008). Polar regions are also prone to the halogen induced catalytic depletion of O_3 (Koo et al. 2012). Barrie et al. (1988) showed that O_3 depleting events in the Arctic boundary layer were associated with high particulate bromine concentrations. Recent observational evidences show that halogen chemistry plays a significant role in the troposphere and is required to be represented in chemistry transport models for an accurate production of present day O_3 budget (Sherwen et al. 2017). Using the GEOS-Chem model of chemistry and transport the continuously increasing significance of halogen chemistry was investigated. The increased halogen content in the troposphere can be attributed to increased oceanic iodine emissions, higher anthropogenic emissions of bromo-carbons and an increased flux of bromine from stratosphere (Sherwan et al. 2017).

The marine ecosystems (oceans) are the important generations of tropospheric halogens (Sarwar et al. 2015). However, a few other sources also play an important role in determining halogen contents in the troposphere (Simpson et al. 2015). Figure 2.4 depicts a schematic representation of the various halogen sources and their role in tropospheric O_3 depletion. O_3 deposition over sea water is highly variable due to difference in atmospheric O_3 mixing ratio and deposition velocity (flux/mixing ratio) (Helmig et al. 2012). Studies have shown that several compounds in seawater can interact with O_3 (Shaw and Carpenter 2013) and the presence of other compounds can influence these interactions (Resser and Donaldson 2011).

However, the measurements done by advanced instrumentations in the past few years have confirmed that oceans are the important source of halogen emissions, mainly bromine and iodine (Prados-Roman et al. 2015; Simpson et al. 2015; Carpenter et al. 2013; Mahajan et al. 2012). With the advent of studies of tropospheric halogen chemistry, various marine emission sources such as halocarbons, sea spray aerosol, bromine, iodine/hypoiodous acid and chemical reactions such as photolysis of higher iodine oxides have been implemented in global chemical transport models (Ordonez et al. 2012; Saiz-Lopez et al. 2014). Saiz-Lopez et al. (2012) found that tropospheric O_3 loss due to marine halogen emission can have a climatically relevant radiative impact whereas Parrella et al. (2012) observed that inclusion of bromine chemistry in the GEOS-Chemical transport model (CTM) improves the model predictions of preindustrial O_3 observation. It has been observed that bromine activated from sea spray aerosols reduces the marine boundary layer O_3 by >20% over most of the tropics and >75% over the Southern Ocean (Long et al. 2014). The reaction between O_3 and dissolved iodide in seawaters not only decrease atmospheric O_3 but also generates halogen emissions that further stimulate halogen induced O_3 depletion in troposphere (Sarwar et al. 2015).

4 Effect of Climate Change on Ozone Budget in Troposphere

Troposphere O_3 budget is largely influenced by the phenomenon called 'climate change' which refers to changes in the global or regional climate patterns evaluated over multidecadal time scale and that persists for extended period of time. Climate change phenomenon is characterized by changes in the atmospheric burden of specific greenhouse gases which leads to changes in radiation budgets, changes in temperature and precipitation patterns, increased frequency of tropical and extratropical storms, and alteration in global air circulation patterns and variation in stratospheric-troposphere circulations (Hartmann et al. 2013). Most of the atmospheric variables that are disturbed/ altered by the present climate change scenario are actually the important components of O_3 formation in the troposphere. It has been estimated that total amount of troposphere O_3 has increased by 30% globally between 1750 and 2000 which largely attributed to climate change phenomenon (Houghtou et al. 2001). Climate change, therefore influences the troposphere O_3 budget by affecting the following components of O_3 formation process:

(i)   Changes in meteorological variables
(ii) Rate of change in the emissions of O_3 precursors
(iii) Variations in stratospheric - tropospheric intrusion of O_3

Studies on present climate change scenario indicates that the global mean surface temperature has increased since 19th century, by 0.85 (0.65 − 1.06)°C over a period 1880–2012 (Hartmann et al. 2013). Increased temperature around the globe has resulted in an increase in tropospheric air humidity since 1970s and increase the heavy precipitation events over land since 1950s. Tropospheric water vapors play an important role in regulating the aerosol contents of the atmospheric and in cloud formation which indirectly affects the incident solar radiations reaching the Earth's surface (Hartmann et al. 2013). Change in climate may also affect the meteorological transport processes and hence alter the intercontinental export and import of the pollutants (Doherty et al. 2013). A climate penalty (O_3 increase due to climate change) of 2–5 ppbv in summer daily maximum 8 h O_3 was recorded in north east United States for climate change in 2050 as compared to 2000s following the SRES A1B scenario (Wu et al. 2008). This increase of O_3 concentration is attributed to a number of meteorological variables including temperature (a rise of 1–3 °C) and mid latitude cyclone frequency (Wu et al. 2008). Changes in temperature and water vapour alter the chemical environment and therefore affect the rate of chemical reactions that create and remove O_3 (Fiore et al. 2012; Jacob and Winner 2009). Many chemical reaction rates increase with temperature e.g. methane and non-methane hydrocarbon oxidation rates leading to increased O_3 production. Increased water vapour in warmer atmosphere in future will lead to increased O_3 destruction and result in shorter O_3 lifetimes (Johnson et al. 1999). Modeling studies have shown that the influence of water vapour has led to a reduction in contribution of Asian emission to background O_3 over United States (Lin et al. 2008).

One of the most important components of O_3 formation that is significantly influenced by climate change is the emission of biogenic VOCs. A rise in temperature exponentially increases the rate of emission of BVOCs (Guenther et al. 2012) either by enhancing the activities of the enzymes involved in its synthesis or by raising the BVOCs vapour pressure by thereby decreasing the resistance of the diffusion pathway (Penuelas & Staudt 2010). It has been estimated that over the past 30 years the BVOC emission has increased by 10% due to the climate warming and an additional 2 − 3°C rise in temperature would further increase the global BVOCs emission by 30–45% (Penuelas and Staudt 2010). In addition to this, warming climate signifies an extended plant activity season in boreal and temperate environments which suggests further increased BVOCs emissions (Penuelas and Filella 2001). Climate change scenario also predicts a change in the precipitation patterns such that some areas are expected to become drier (IPCC 2007). Reduced precipitation promotes drought events and subsequent heat stress in plants which affect the BVOC emissions (Penuelas and Filella 2001).

Atmospheric lightning process which is considered to be a dominant source of NOx in the upper troposphere, shows a linear response with respect to changes in global surface temperature (Finney et al. 2016). NOx produced in the upper tropospheric layer is more efficient at catalyzing O_3 production than surface emissions (Dahlmann et al. 2011). Lighting NOx is found to have an O_3 production efficiency of 6.5 ± 4.7 times that of surface NOx source (Finney et al. 2016). Estimates from climate chemistry models suggest that lightning NOx emission will increase by 4–60% per degree increase in global mean surface temperature (Banerjee et al. 2014; Zeng et al. 2008). Banerjee et al. (2014) working with United Kingdom Chemistry and Aerosol submodel (UKCA) investigated the effect of NOx emission from lightning on O_3 production and observed that O_3 production increased linearly with NOx emissions from lightning. Based upon this modeling study, tropospheric O_3 burden in response to lightning NOx emissions increased by $413 \pm 28\,Tg\,(O_3)\,yr^{-1}$ and $979 \pm 33\,Tg\,(O_3)\,yr^{-1}$ for the RCP 4.5 and RCP 8.5 scenarios respectively (Banerjee et al. 2014). In addition to this, the lightning emitted NOx also alters the oxidizing capacity of the troposphere (Finney et al. 2016; Banerjee et al. 2014). For instance the chemistry of HOx family can be altered by the conversion of $HO_2°$ to $OH°$ via the reaction between HO_2 and NO. On the other hand HOx loss can be stimulated by the reaction of $OH°$ and NO_2 to form nitric acid (Schumann and Huntrieser 2007). Alteration in HOx concentration in the troposphere indirectly affects the lifetimes of methane via its reaction with the $OH°$ radicals (Holmes et al. 2013; Murray et al. 2014).

As discussed earlier, the STE events also contribute a fair share in the tropospheric O_3 budget (Hofmann et al. 2016; Lefohn et al. 2012). The Brewer Dobson circulation which plays a significant role in STE events is greatly influenced by the changing climatic conditions (Bunzel and Schmidt 2013). Modeling studies have shown that greenhouse gas induced climate change stimulated an ~ 2–3.2% per decade acceleration of global mass circulation of tropospheric air through the stratosphere (Butchart 2014). Karpechko and Manzini (2012) predicted a strengthening of Brewer – Dobson circulation under increasing greenhouse gas concentration, thus increasing the STE events. Change in Brewer Dobson circulation brings about changes in the stratospheric O_3 composition by modifying the transport of O_3 itself and that of ozone depleting substances such as chlorofluorocarbons (CFCs) (Bunzel and Schmidt 2013). Jiang et al. (2007) observed that the strengthening of Brewer Dobson circulation results in a decline in O_3 concentrations in tropics and an increase in O_3 concentrations in high latitudes. It has been observed that the mechanism responsible for strengthening the Brewer Dobson circulation is attributed to the increase in the concentration of greenhouse gases which increase the temperature of tropospheric and decrease the stratospheric temperature (Garcia and Randel 2008). Stolarski et al. (2015) also emphasized the role of greenhouse gases specifically CO_2 in lowering the stratospheric temperature thus affecting the distribution of stratospheric O_3. CO_2 cools the stratosphere and slows down the temperature dependant O_3 loss processes, resulting in rising stratospheric O_3 levels (Goessling and Bathiany 2016). In addition to this, CO_2 also speeds up the Brewer-Dobson circulation (Butchart 2014). The combination of stratospheric cooling and speed- up of Brewer Dobson circulation results in tremendous O_3 recovery in the stratosphere which stimulates the frequency of STE events (Zeng et al. 2010; Banerjee et al. 2016) (as discussed in the earlier section of the chapter). Stratospheric cooling due to enhanced CO_2 also assists in alleviating O_3 loss in the stratosphere due to N_2O (Stolarski et al. 2015). Several studies have examined the interactive effect of CO_2 and N_2O on stratospheric O_3 using two dimensional (Portmann et al. 2012) or three dimensional (Revell et al. 2012a, b; Wang et al. 2014) models and have shown that CO_2 induced stratospheric cooling and strengthening of Brewer Dobson circulation reduced the O_3 depleting role of N_2O. Based upon the above mentioned modeling studies it can be predicted that under future climate change scenario, role of NOx will have only a small effect on stratospheric O_3 budget (Oman et al. 2010). Using a two dimensional Goddard space Flight Centre coupled chemistry- radiations- dynamics model (GSFC2D) a small increase (~5 DU) in stratospheric O_3 column was observed in 2100 as compared to that in 1960 for constant methane concentration (Stolarski et al. 2015). This increase in the O_3 column is attributed to the projected increase in CO_2, conferring the role of this greenhouse gas in stratospheric O_3 recovery and subsequent promotion of STE event frequency (Stolarski et al. 2015).

5 Conclusion

Tropospheric O_3 is a climate gas and a potential secondary air pollutant. Estimating the changes in O_3 concentration from natural pre-industrial concentration to the present day concentration help in formulating policies for maintaining the air quality. Important drivers affecting the tropospheric O_3 chemistry are the meteorological variables such as temperature and incident solar radiations which show a positive relationship with O_3 concentration. Maximum temperature directly affects the different processes of O_3 chemistry and assumes the role of the key driver increasing O_3 concentration which often exceeds the air quality standards. Relative humidity is another important meteorological variable in determining the O_3 concentration in a region particularly during the spring time.

The meteorological variables, in addition to directly affecting the tropospheric O_3 chemistry also influences a few physical and physiological processes involved in O_3 formation. For instance, high temperature not only affects the lightening process resulting in increased NOx concentration in the upper tropospheric layer, it also promotes the BVOC emissions which are directly involved in O_3 formation. Climate change brings about certain alterations in the atmospheric events such as warmer temperatures, altered air circulations, increased heat waves etc. Increasing CO_2 concentrations in the present climate change scenario increases the frequency of stratospheric tropospheric exchange events that predict a further increase in tropospheric O_3 budget. Halogen induced depletion of O_3 in troposphere also plays an important role in maintaining O_3 budget in the troposphere. Predominant in marine and polar regions, halogen chemistry results in a significant reduction of global O_3 concentration.

References

Abbatt JPD, Lee AKY, Thornton JA (2012) Quantifyingtrace gas uptake to tropospheric aerosol: recent advances and remaining challenges. Chem Soc Rev 41:6555–6581

Alves EG, Peter Harley P, Goncalves JFC, Moura CES, Jardine K (2014) Effects of light and temperature on isoprene emission at different leaf developmental stages of *Eschweileracoriacea* in central Amazon. Acta Amazon 44(1):9–18

Ambrose JL, Reidmiller DR, Jaffe DA (2011) Causes of high O_3 in the lower freetroposphere over the Pacific Northwest as observed at the Mt. Bachelor Observatory. Atmos Environ 45:5302–5315

Andersson C, Engardt M (2010) European ozone in a future climate: importance of changes in dry deposition and isoprene emissions. J Geophys Res Atmos 115:D02303

Arneth A, Schurgers G, Lathiere J, Duhl T, Beerling DJ, Hewitt CN, Martin M, Guenther A (2011) Global terrestrial isoprene emission models: sensitivity to variability in climate and vegetation. Atmos Chem Phys 11:8037–8052

Aumont B, Chervier F, Laval S (2003) Contribution of HONO sources to the $NOx/HOx/O_3$ chemistry in the polluted boundary layer. Atmos Environ 37(4):487–498

Barrie LA, Bottenheim JW, Schnell RC, Crutzen PJ, Rasmussen RA (1988) Ozone destruction and photochemical reactions at polar sunrise in lower arctic atmosphere. Nature 334:138–141

Banerjee A, Archibald AT, Maycock AC, Telford P, Abraham NL, Yang X, Braesicke P, Pyle JA (2014) Lightning NOx, a key chemistry-climate interaction: impacts of future climate change and consequences for tropospheric oxidising capacity. Atmos Chem Phys 14:9871–9881

Banerjee A, Maycock AC, Archibald AT, Abraham NL, Paul Telford P, Braesicke P, Pyle JA (2016) Drivers of changes in stratospheric and tropospheric ozone between year 2000 and 2100. Atmos Chem Phys 16:2727–2746

Berezina EV, Moiseenko KB, Skorokhod AI, Corresponding Member of the RAS Elansky NF, Belikov IB (2017) Aromatic volatile organic compounds and their role in ground-level ozone formation in Russia. Dokl Akad Nauk 474(3):356–360

Bezuglaya EY, Smirnova IV (2008) Cities' Air and Its Variations (Asterion, St. Petersburg, 2008) [in Russian]

Bourqui MS, Trepanier PY (2010) Descent of deep stratospheric intrusions during the IONS August 2006 campaign. J Geophys Res 115:D18–301

Bunzel F, Schmidt H (2013) The Brewer–Dobson circulation in a changing climate: impact of the model configuration. J Atmos Sci 70:1437–1455

Butchart N (2014) The Brewer–Dobson circulation. Rev Geophys 52:157–184

Carr S, Heard DE, Blitz MA (2009) Comment on Atmospheric Hydroxyl Radical Production from Electronically Excited NO2 and H2O. Science 324(5925):336

Carslaw DC, Beevers SD, Tate JE, Westmoreland EJ, Williams ML (2011) Recent evidence concerning higher NOx emissions from passenger cars and light duty vehicles. Atmos Environ 45:7053–7063

Carpenter LJ, MacDonald SM, Shaw MD, Kumar R, Saunders RW, Parthipan R, Wilson J, Plane JMC (2013) Atmospheric iodine levels influenced by sea surface emissions of inorganic iodine. Nat Geosci 6(2):108–111

Chameides WL, Walker JCG (1973) A photochemical theory fortropospheric ozone. J Geophys Res 78:8751–8760

Chatani S, Amann M, Goel A, Hao J, Klimont Z, Kumar A, Mishra A, Sharma S, Wang SX, Wang YX, Zhao B (2014) Photochemical roles of rapid economic growth and potential abatement strategies on tropospheric ozone over South and East Asia in 2030. Atmos Chem Phys 14:9259–9277

Chatfield R, Harrison H (1976) Ozone in the remote troposphere –mixing versus photochemistry. J Geophys Res-Ocean Atmos 81:421–423

Chen X, Anel JA, Su Z, de la Torre L, Kelder H, van Peet J, Ma Y (2013) The deep atmospheric boundary layerand its significance to the stratosphere and troposphere exchange over the Tibetan Plateau. PLoS One 8:e56909

Cofala J, Amann M, Klimont Z, Kupiainen K, Hoglund-Isaksson L (2007) Scenarios of global anthropogenic emissions of air pollutants and methane until 2030. Atmos Environ 41(38):8486–8499. https://doi.org/10.1016/j.atmosenv.2007.07.010

Coates J, Mar KA, Ojha N, Butler TM (2016) The influence of temperature on ozone production under varying NOx conditions – a modelling study. Atmos Chem Phys 16:11601–11615

Collins WJ, Derwent RG, Garnier B, Johnson CE, Sanderson MG, Stevenson DS (2003) Effect of stratosphere troposphere exchange on the future tropospheric ozone trend. J Geophys Res 108:8528. https://doi.org/10.1029/2002JD002617

Cooper OR, Oltmans SJ, Johnson BJ, Brioude J, Angevine W, Trainer M, Parrish DD, Ryerson TR, Pollack I, Cullis PD, Ives MA, Tarasick DW, Al-Saadi J, Stajner I (2011) Measurement of western U.S. baseline ozone from the surface to the tropopause and assessment of downwind impact regions. J Geophys Res 11:D00V03

Crutzen PJ (1973) Photochemical reactions initiated by and influencing ozone in the unpolluted troposphere. Tellus 26:47–57

Dalsøren SB, Eide MS, Myhre G, Endresen O, Isaksen ISA, Fuglestvedt AS (2010) Impacts of the Large Increase in International Ship Traffic 2000–2007 on Tropospheric Ozone and Methane. Environ Sci Technol 44:2482–2489

Dahlmann K, Grewe V, Ponater M, Matthes S (2011) Quantifying the contributions of individual NOx sources to the trend in ozone radiative forcing. Atmos Environ 45(17):2860–2868

Danielsen EF (1968) Stratospheric-tropospheric exchange based on radioactivity, ozone and potential vorticity. J Atmos Sci 25:502–518

Demuzere M, Trigo RM, Vila-Guerau de Arellano J, van Lipzig NPM (2009) The impact of weather and atmospheric circulation on O3 and PM10 levels at a rural mid latitude site. Atmos Chem Phys 9:2695–2714

Derwent R, Simmonds P, Manning A, Spain T (2007) Trends overa 20-year period from 1987 to 2007 in surface ozone at the atmosphericresearch station, Mace Head, Ireland. Atmos Environ 41:9091–9098

Doherty RM et al (2013) Impacts of climate change on surface ozone and intercontinental ozone pollution: a multi-model study. J Geophys Res Atmos 118:3744–3763

Dreyfus GB, Schade GW, Goldstein AH (2002) Observational constraints on the contribution of isoprene oxidation to ozone production on the western slope of the Sierra Nevada, California. J Geophys Res 107(D19):4365. GAW report no. 205, Impacts of megacities on air pollutionand climate (World Meteor. Org., Geneva, 2012)

Ehn M, Thornton JA, Kleist E, Sipilä M, Junninen H, Pullinen I et al (2014) A large source of low-volatility secondary organic aerosol. Nature 506(7489):476–479

Elshorbany Y, Barnes I, Becker KH, Kleffmann J, Wiesen P (2010) Sources and cycling of tropospheric hydroxyl radicals –an overview. Z Phys Chem 224(7–8):967–987

Environmental Protection Agency (2007) Air Quality and Emissions – Progress Continues in 2006. http://www.epa.gov/airtrends/econ-emissions.html

Escudero M, Lozano A, Hierro J, Valle JD, Mantilla E (2014) Urban influence on increasing ozone concentrations in a characteristic Mediterranean agglomeration. Atmos Environ 99:322–332

Eyring V, Isaksen ISA, Berntsen T, Collins WJ, Corbett JJ, Endresen Ø, Grainger RG, Moldanova J, Schlager H, Stevenson DS (2010) Assessment of transport impacts on climate and ozone: shipping. Atmos Environ 44:4735–4771

Finney DL, Doherty RM, Wild O, Abraham NL (2016) The impact of lightning on tropospheric ozone chemistry using a new global parametrisation. Atmos Chem Phys Discuss 16:1–28

Fiore AM et al (2012) Global air quality and climate. Chem Soc Rev 41(19):6663–6683

Fu Y, Tai APK (2015) Impact of climate and land cover changes on tropospheric ozone air quality and public health in East Asia between 1980 and 2010. Atmos Chem Phys 15:10093–10106

Garcia RR, Randel WJ (2008) Acceleration of the Brewer– Dobson circulation due to increases in greenhouse gases. J Atmos Sci 65:2731–2739

Gligorovski S, Strekowski R, Barbati S, Vione D (2015) Environmental Implications of Hydroxyl Radicals (•OH). Chem Rev 115:13051–13092

Goessling HF, Bathiany S (2016) Why CO2 cools the middle atmosphere-a consolidating model perspective. Earth System Dyn 7:697–715

Guenther A, Hewitt C, Erickson D, Fall R, Geron C, Graedel T, Harley P, Klinger L, Lerdau M, McKay W, Pierce T, Scholes R, Steinbrecher R, Tallamraju R, Taylor J, Zimmerman P (1995) A global model of natural volatile organic compound emissions. J Geophys Res 100:8873–8892

Guenther AB, Jiang X, Heald CL, Sakulyanontvittaya T, Duhl T, Emmons LK, andWang X (2012) The model of emissions of gases and aerosols from nature version 2.1 (MEGAN2.1): an extended and updated framework for modeling biogenic emissions. Geosci Model Dev 5:1471–1492

Haagen-Smit AJ (1952) Chemistry and physiology of Los Angeles smog. Ind Eng Chem 44:1342

Hartmann DL, Klein Tank AMG, Rusticucci M, Alexander L, Brönnimann S, Charabi Y, Dentener F, Dlugokencky E, Easterling D, Kaplan A, Soden B, Thorne P, Wild M, Zhai PM (2013) Observations: Atmosphere and surface supplementary material. In: Stocker TF, Qin D, Plattner G-K, Tignor M, Allen SK, Boschung J, Nauels A, Xia Y, Bex V, Midgley PM (eds) Climate change 2013: the physical science basis. Contribution of working group I to the fifth assessment report of the Intergovernmental Panel on climate change. Cambridge University Press, Cambridge, US

Hauglustaine DA, Koffi B (2012) Boundary layer ozone pollution caused by future aircraft emissions. Geophys Res Lett 39:L13808

Hegglin M, Shepherd T (2009) Large climate-induced changes in ultraviolet index and stratosphere-to-troposphere ozone flux. Nat Geosci 2:687–691

Helmig D, Boylan P, Johnson B, Oltmans S, Fairall C et al (2012) Ozone dynamics and snow-atmosphere exchanges during ozone depletion events at Barrow, Alaska. J Geophys Res Atmos 117(D20). ISSN 01480227. https://doi.org/10.1029/2012JD017531

Hess PG, Zbinden R (2013) Stratospheric impact on tropospheric ozone variability and trends: 1990–2009. Atmos Chem Phys 13:649–674

Hofmann C, Kerkweg A, Hoor P, Jöcke P (2016) Stratosphere-troposphere exchange in the vicinity of a tropopause fold. Atmos Chem Phys Discuss 2015:949–975

Holmes CD, Prather MJ, Sovde OA, Myhre G (2013) Future methane, hydroxyl and their uncertainties: key climate and emission parameters for future predictions. Atmos Chem Phys 13:285–302

Ibrahim AE-DMM (2014) Nox and Sox emissions and climate changes. World Appl Sci J 31(8):1422–1426

IPCC (2007) The physical science basis. contribution of working group I. In: Solomon S et al (eds) Fourth assessment report of the intergovernmental panel on climate change. Cambridge University Press, Cambridge, UK, pp 1–996

IPCC (2013) In: Stocker TF, Qin D, Plattner G-K, Tignor M, Allen SK, Boschung J, Nauels A, Xia Y, Bex V, Midgley PM (eds) Climate Change 2013: The Physical Science Basis. Contribution of Working Group I to the Fifth Assessment Report of the Intergovernmental Panel on Climate Change. Cambridge University Press, Cambridge, UK\New York, NY

Isaksen I, Berntsen T, Dalsøren S, Eleftheratos K, Orsolini Y, Rognerud B, Stordal F, Søvde O, Zerefos C, Holmes C (2014) Atmospheric ozone and methane in a changing climate. Atmosphere 5:518–535

Jacob DJ, Winner DA (2009) Effect of climatechange on air quality. Atmos Environ 43:51–63

Jacobson MZ, Streets DG (2009) Influence of future anthropogenic emissions on climate, natural emissions, and air quality. J Geophys Res-Atmos 114:D08118

Jenkinson AF, Collison FP (1977) An initial climatology of gales over the North Sea Synoptic. Climatol Branch Memorandum 62:18. UK Met Office

Jiang X, Ku W, Shia R, Li Q, Elkins J, Prinn R, Yung Y (2007) Seasonal cycle of N_2O: analysis of data. Global Biogeochem Cycles 21:GB1006

Jöckel P, Kerkweg A, Pozzer A, Sander R, Tost H, Riede H, Baumgaertner A, Gromov S, Kern B (2010) Development cycle 2 of the modular earth submodel system (MESSy2). Geosci Model Dev 3:717–752

Johnson CE, Collins WJ, Stevenson DS, Derwent RG (1999) Relative roles of climate and emissions changes on future oxidant concentrations. J Geophys Res 104(D15):18631–18645

Karpechko AY, Manzini E (2012) Stratospheric influence ontropospheric climate change in the Northern Hemisphere. J Geophys Res 117:D05133

Kawase H, Nagashima T, Sudo K, Nozawa T (2011) Future changes in tropospheric ozone under representative concentration pathways (RCPs). Geophys Res Lett 38:L05801

Kerkweg A, Jöckel P (2012) The 1-way on-line coupled atmospheric chemistry model system MECO(n) - Part 1: Description of the limited area atmospheric chemistry model COSMO/MESSy. Geosci Model Dev 5:87–110

Kim S, VandenBoer TC, Young CJ, Riedel TP, Thornton JA, Swarthout B, Sive B, Lerner B, Gilman JB, Warneke C, Roberts JM, Guenther A, Wagner NL, Dubé WP, Williams E, Brown SS (2014) The primary and recycling sources of OH during the NACHTT-2011 campaign: HONO as an important OH primary source in the wintertime. J Geophys Res-Atmos 119:JD019784

Kleffmann J, Gavriloaiei T, Hofzumahaus A, Holland F, Koppmann R, Rupp L, Schlosser E, Siese M, Wahner A (2005) Day time formation of nitrous acid: a major source of OH radicals in a forest. Geophys Res Lett 32(5):L05818

Knowlton K, Rosenthal JE, Hogrefe C, Lynn B, Gaffin S, Goldberg R, Rosenzweig C, Civerolo K, Ku JY, Kinney PL (2004) Assessingozone-related health impacts under a changing climate. Environ Health Perspect 112(15):1557–1563

Koo J-H, Wang Y, Kurosu TP, Chance K, Rozanoy A, Richter A, Oltmans SJ, Thompson AM, Hair JW, Fenn MA, Weinheimer AJ, Ryerson TB, Solberg S, Huey LG, Liao J, Dibb JE, Neuman JA, Nowak JB, Pierce RB, Natarajan M, Al-Saadi J (2012) Characteristics of tropospheric ozone depletion events in the Arctic spring: analysis of the ARCTAS, ARCPAC, and ARCIONS measurements and satellite BrO observations. Atmos Chem Phys 12:9909–9922

Kurokawa J, Ohara T, Morikawa T, Hanayama S, Janssens- Maenhout G, Fukui T, Kawashima K, Akimoto H (2013) Emissions of air pollutants and greenhouse gases over Asian regions during 2000–2008: regional emission inventory in Asia (REAS) version 2. Atmos Chem Phys 13:11019–11058

Lamarque J-F, Kyle GP, Meinshausen M, Riahi K, Smith SJ, van Vuuren DP, Conley AJ, Vitt F (2011) Globaland regional evolution of short-lived radiatively-active gases and aerosols in the representative concentration pathways. Clim Chang 109:191–212

Langford AO, Brioude J, Cooper OR, Senff CJ, Alvarez RJ II, Hardesty RM, Johnson BJ, Oltmans SJ (2012) Stratospheric influence on surface ozone in the Los Angeles area during late spring and early summer of 2010. J Geophys Res 117:D00V06

Lefohn AS, Wernli H, Shadwick D, Limbach S, Oltmans SJ, Shapiro M (2011) The importance of stratospheric-tropospheric transport in affecting surface ozone concentrations in the western and northern tier of the United States. Atmos Environ 45:4845–4857. https://doi.org/10.1016/j.atmosenv.2011.06.014

Lefohn AS, Wernli H, Shadwick D, Oltmans SJ, Shapiro M (2012) Quantifying the importance of stratospheric-tropospheric transport on surfaceozone concentrations at high- and low-elevation monitoring sites in the United States. Atmos Environ 62:646–656

Leighton PA (1961) Photochemistry of air pollution. Academic Press, New York

Lin J-T et al (2008) Effects of intercontinental transport on surface ozone over the United States: present and future assessment with a global model. Geophys Res Lett 35:L02805

Lin MY, Fiore AM, Cooper OR, Horowitz LW, Langford AO, Levy H II, Johnson BJ, Naik V, Oltmans SJ, Senff CJ (2012) Springtime high surface ozone events over the western United States: quantifying the roleof stratospheric intrusions. J Geophys Res 117:D00V22

Lindwall F, Faubert P, Rinnan R (2015) Diel variationof biogenic volatile Organic compound emissions- a field study in the sub, low and high arctic on the effect of temperature and light. PLoS One 10(4):e0123610

Logan JA, Prather MJ, Wofsy SC, McElroy MB (1981) Tropospheric chemistry: a global perspective. J Geophys Res 86:7210

Logan JA, Staehelin J, Megretskaia IA, Cammas JP, Thouret V, Claude H, De Backer H, Steinbacher M, Scheel HE, Stubi R, Frohlich M, Derwent R (2012) Changes in ozone over Europe: analysis of ozone measurements from sondes, regularaircraft (MOZAIC) and alpine surface sites. J GeophysRes-Atmos 117:D09301

Long MS, Keene WC, Easter RC, Sander R, Liu X, Kerkweg A, Erickson D (2014) Sensitivity of tropospheric chemicalcomposition to halogen-radical chemistry using a fully coupledsize-resolved multiphase chemistry–global climate system:halogen distributions, aerosol composition, and sensitivity ofclimate-relevant gases. Atmos Chem Phys 14:3397–3425

Loothawornkitkul J, Taylor JE, Paul ND, Hewitt CN (2009) Biogenic volatile organic compound in earth system. New Phytol 183(1):27–51

Ma JZ, Yang LX, Shen XL, Qin JH, Deng LL, Ahmed S, Xu HX, Xue DY, Ye JX, Xu G (2014) Effects of traditional Chinese medicinal plants on antiinsulin resistance bioactivity of DXMS-induced insulin resistant HepG2 cells. Nat Prod Bioprospect 4:197–206

Mahajan AS, Gómez Martín JC, Hay TD, Royer S-J, Yvon-Lewis S, Liu Y, Hu L, Prados-Roman C, Ordóñez C, Plane JMC, Saiz-Lopez A (2012) Latitudinal distribution of reactive iodine in the Eastern Pacific and its link to open ocean sources. Atmos Chem Phys 12:11609–11617. https://doi.org/10.5194/acp-12-11609-2012

McKinney KA, Lee BH, Vasta A, Pho TV, Munger JW (2011) Emissions of isoprenoids and oxygenated biogenicvolatile organic compounds from a New England mixed forest. Atmos Chem Phys 11(10):4807–4831

Monks PS (2005) Gas-phase radical chemistry in the troposphere. Chem Soc Rev 34:376–395

Monks PS, Archibald AT, Colette A, Cooper O, Coyle M, Derwent R, Fowler D, Granier C, Law KS, Mills GE, Stevenson DS, Tarasova O, Thouret V, von Schneidemesser E, Sommariva R, Wild O, Williams ML (2015) Tropshericozone and its precursors from the urban to the global scale fromair quality to short-lived climate forcer. Atmos Chem Phys 15:8889–8973

Moore G, Semple JL (2005) A Tibetan Taylor cap and a halo of stratospheric ozone over the Himalaya. Geophys Res Lett 32:L21810

Murray LT, Mickley LJ, Kaplan JO, Sofen ED, Pfeiffer M, Alexander B (2014) Factors controlling variability in the oxidative capacity of the troposphere since the last Glacial maximum. Atmos Chem Phys 14:3589–3622

Nakicenovic N, Alcamo J, Davis G, de Vries B, Fenhann J, Gaffin S, Gregory K, Grubler A, Jung TY, Kram T, La Rovere EL, Michaelis L, Mori S, Morita T, Pepper W, Pitcher HM, Price L, Riahi K, Roehrl A, Rogner H-H, Sankovski A, Schlesinger M, Shukla P, Smith SJ, Swart R, van Rooijen S, Victor N, Dadi Z (2000) Special report on Emissions scenarios: a special report of Working Group III of the Intergovernmental Panel on Climate Change., Other Information: PBD: 03 Octo 2000. Cambridge University Press, New York

Neu JL, Flury T, Manney GL, Santee ML, Livesey NJ, Worden J (2014) Tropospheric ozone variations governed by changes in stratospheric circulation. Nat Geosci 7:340–344

Olsen MA, Douglass AR, Kaplan TB (2013) Variability of extra tropical ozone stratosphere-troposphere exchange using microwave limb sounder observations. J Geophys Res-Atmos 118:1090–1099

Oltmans SJ, Lefohn AS, Shadwick D, Harris JM, Scheel HE, Galbally I, Tarasick DW, Johnson BJ, Brunke EG, Claude H, Zeng G, Nichol S, Schmidlin F, Davies J, Cuevas E, Redondas A, Naoe H, Nakano T, Kawasato T (2013) Recent tropospheric ozone changes – a pattern dominated by slow or no growth. Atmos Environ 67:331–351

Oman LD, WaughDW KSR, Stolarski RS, Douglass AR, Newman PA (2010) Mechanisms and feedback causing changes in upper stratospheric ozone in the 21st century. J Geophys Res 115:D05303

Ordoñez C, Brunner D, Staehelin J, Hadjinicolaou P, Pyle JA, Jonas M, Wernli H, Prevot ASH (2007) Strong influence of lowermost stratospheric ozone on lower tropospheric background ozone changes over Europe. Geophys Res Lett 34:L07805

Ordóñez C, Lamarque J-F, Tilmes S, Kinnison DE, Atlas EL, Blake DR, Sousa Santos G, Brasseur G, Saiz-Lopez A (2012) Bromine and iodine chemistry in a global chemistry-climate model: description and evaluation of very short-lived oceanic sources. Atmos Chem Phys 12:1423–1447

Ormeno E, Gentner DR, Fares S, Karlik J, Park JH, Goldstein AH (2010) Environ Sci Technol 44:3758–3764

Otero N, Sillmann J, Schnell JL, Rust HW, Butler T (2016) Synoptic and meteorological drivers of extreme ozone concentrations over Europe. Environ Res Lett 11:024005

Pacifico E, Folberth GA, Jones CD, Harrison SP, Collins WJ (2012) Sensitivity of biogenic isoprene emissions to past, present and future environmental conditions and implications for atmospheric chemistry. J Geophys Res 117:D22302

Parrella JP, Jacob DJ, Liang Q, Zhang Y, Mickley LJ, Miller B, Evans MJ, Yang X, Pyle JA, Theys N, Van Roozendael M (2012) Tropospheric bromine chemistry: implications forpresent and pre-industrial ozone and mercury. Atmos Chem Phys 12:6723–6740

Parrish DD, Law KS, Staehelin J, Derwent R, Cooper OR, Tanimoto H, Volz-Thomas A, Gilge S, Scheel H-E, Steinbacher M, Chan E (2012) Long-term changes in lower troposphericbaseline ozone concentrations at northern mid-latitudes. Atmos Chem Phys 12:11485–11504

Penuelas J, Filella I (2001) Phenology: responses to a warming world. Science 294:793–795

Penuelas J, Staudt M (2010) BVOCs and global change. Trends Plant Sci 15(3):133–144

Peñuelas J, Staudt M (2010) The emission factor of volatile isoprenoids: stress, acclimation, and developmental responses. Biogeosciences 7:2203–2223

Portmann R, Daniel J, Ravishankara A (2012) Stratospheric ozone depletion due to nitrous oxide: influences of other gases. Philos Trans R Soc B 367:1256–1264

Pruchniewicz PG (1977) Meridional distribution of ozonein the troposphere and its seasonal variations. J Geophys Res 82:2063–2073

Prados-Roman C, Cuevas CA, Hay T, Fernandez RP, Mahajan AS, Royer S-J, Galí M, Simó R, Dachs J, Großmann K, Kinnison DE, Lamarque J-F, Saiz-Lopez A (2015) Iodine oxide in the global marine boundary layer. Atmos Chem Phys 15:583–593. https://doi.org/10.5194/acp-15-583-2015

Pusede SE, Gentner DR, Wooldridge PJ, Browne EC, Rollins AW, Min K-E, Russell AR, Thomas J, Zhang L, Brune WH, Henry SB, DiGangi JP, Keutsch FN, Harrold SA, Thornton JA, Beaver MR, St. Clair JM, Wennberg PO, Sanders J, Ren X, VandenBoer TC, Markovic MZ, Guha A, Weber R, Goldstein AH, Cohen RC (2014) On the temperature dependence of organic reactivity, nitrogen oxides, ozone production, and the impact of emission controls in San Joaquin Valley, California Atmos. Chem Phys 14:3373–143395

Pusede SE, Steiner AL, Cohen RC (2015) Temperature and recent trends in the chemistry of Continental surface ozone. Chem Rev 115:3898–3918

Read KA, Majajan AS, Carpenter LJ, Evans MJ, Faria BVE, Heard DE, Hopkins JR, Lee JD, Moller SJ, Lewis AC, Mendes L, McQuaid JB, Oetjen H, Saiz-Lopez A, Pilling MJ, Plane JMC (2008) Extensive halogen mediated ozone destruction over the tropical Atlantic Ocean. Nature 453:1232–1235

Resser DI, Donaldson DJ (2011) Influence of water surface properties on the heterogeneous reaction between O3(g) and I(aq)−. Atmos Environ 45:6116–6120

Reed C, Evans MJ, Crilley LR, Bloss WJ, Sherwen T, Read KA, Lee JD, Carpenter LJ (2017) Evidence for renoxification in the tropical marine boundary layer. Atmos Chem Phys 17:4081–4092

Reutter P, Škerlak B, Sprenger M, Wernli H (2015) Stratosphere–troposphere exchange (STE) in the vicinity of North Atlantic cyclones, Atmos. Chem Phys 15(10):939–10 953

Revell L, Bodeker G, Huck P, Williamson B, Rozanov E (2012a) The sensitivity of stratospheric ozone changes through the 21st century to N2O and CH4 Atmos. Chem Phys 12:309–311

Revell L, Bodeker G, Smale D, Lehmann R, Huck P, Williamson B, Rozanov E, Struthers H (2012b) The effectiveness of N2O in depleting stratospheric ozone. Geophys Res Lett 39:1–6

Revell LE, Tummon F, Stenke A, Sukhodolov T, Coulon A, Rozanov E, Garny H, Grewe V, and-Peter T (2015) Drivers of the tropospheric ozone budget throughout the 21st century under the medium-high climate scenario RCP 6.0. Atmos Chem Phys 15:5887–5902

Rohrer F, Berresheim H (2006) Strong correlation between levels of tropospheric hydroxyl radicals and solar ultraviolet radiation. Nature 442(7099):184–187

Royal Society (2008) Ground-level ozone in the 21st century: futuretrends, impacts and policy implications. The Royal Society, London

Saavedra S, Rodríguez A, Taboada JJ, Souto J, AandCasares JJ (2012) Synoptic patterns and air mass transport during ozoneepisodes in northwestern Iberia. ScienceTotal Environ 441:97–110

Saiz-Lopez A, von Glasow R (2012) Reactive halogen chemistryin the troposphere. Chem Soc Rev 41:6448–6472

Saiz-Lopez A, Lamarque J-F, Kinnison DE, Tilmes S, Ordóñez C, Orlando JJ, Conley AJ, Plane JMC, Mahajan AS, Sousa Santos G, Atlas EL, Blake DR, Sander SP, Schauffler S, Thompson AM, Brasseur G (2012) Estimating the climate significance of halogen-driven ozone loss in the tropical marine troposphere. Atmos Chem Phys 12:3939–3949

Saiz-Lopez A, Fernandez RP, Ordóñez C, Kinnison DE, Gómez Martín JC, Lamarque J-F, Tilmes S (2014) Iodine chemistry in the troposphere and its effect on ozone. Atmos Chem Phys 14:13119–13143. https://doi.org/10.5194/acp-14-13119-2014

Sarwar G, Gantt B, Schwede D, Foley K, Mathur, RandSaiz-Lopez A (2015) Impact of Enhanced ozone deposition and halogen chemistry on tropospheric ozone over the Northern Hemisphere. Environ Sci Technol 49:9203–9211

Schmidt JA, Jacob DJ, Horowitz HM, Hu L, Sherwen T, Evans MJ, Liang Q, Suleiman RM, Oram DE, Le Breton M, Percival CJ, Wang S, Dix B, Volkamer R (2016) Modeling the observed tropospheric BrO background: importance of multiphase chemistry and implications for ozone, OH, and mercury. GeophysRes Atmos *121*:11819–11835

Schollert M, Burchard S, Faubert P, Michelsen A, Rinnan R (2014) Biogenic volatile organic compound emissionsin four vegetation types in high arctic Greenland. Polar Biol 37(2):237–249

Schumann U, Huntrieser H (2007) The global lightning-induced nitrogen oxides source. *Atmos Chem Phys 7*:3823–3907

Schuur EAG, McGuire AD, Schadel C, Grosse G, Harden JW, Hayes DJ, Hugelius G, Koven CD, Kuhry P, Lawrence DM, Natali SM, Olefeldt D, Romanovsky VE, Schaefer K, Turetsky MR, Treat CC, Vonk JE (2015) Climate change and permafrost carbon feedback. Nature 520:171–179

Shakhova NE, Sergienko VI, Semiletov IP (2009) The contribution of the East Siberian shelf to the modern methane cycle. Her Russ Acad Sci *79*:217–246

Shakhova NE, Alekseev VA, Semiletov IP (2010) Predicted methane emission on the east Siberian shelf. *Dokl Earth Sci 430*:190–193

Sherwen T, Schmidt JA, Evans MJ, Carpenter LJ, Großmann K, Eastham SD, Jacob DJ, Dix B, Koenig TK, Sinreich R, Ortega I, Volkamer R, Saiz-Lopez A, Prados- Roman C, Mahajan AS, Ordóñez C (2016) Global impacts of tropospheric halogens (Cl, Br, I) on oxidants and composition in GEOS-Chem. Atmos Chem Phys 16:12239–12271

Sherwen T, Evans MJ, Carpenter LJ, Schmidt JA, andMickley LJ (2017) halogen chemistry reduces tropospheric O3 radiative forcing. Atmos Chem Phys 17:1557–1569

Shindell D, Faluvegi G, Lacis A, Hansen J, Ruedy R, Aguilar E (2006) Role of tropospheric ozone increases in 20th centuryclimate change. J Geophys Res-Atmos 111:L04803

Sicard P, De Marco A, Troussier F, Renou C, Vas N, Paoletti E (2013) Decrease in surface ozone concentrations at Mediterranean remote sites and increase in the cities. Atmos Environ 79:705–715

Simpson WR, Brown SS, Saiz-Lopez A, Thornton JA, von Glasow R (2015) Tropospheric halogen chemistry: sources, cycling, and impacts. Chem Rev 115:4035–4062

Sindelarova K, Granier C, Bouarar I, Guenther A, Tilmes S, Stavrakou T, Müller J-F, Kuhn U, Stefani P, Knorr W (2014) Global data set of biogenic VOC emissions calculated by the MEGAN model over the last 30 years. Atmos Chem Phys 14:9317–9341

Škerlak B, Sprenger M, Wernli H (2014) A global climatology of stratosphere–troposphere exchange using the ERA- Interim data set from 1979 to 2011. Atmos Chem Phys 14:913–937. https://doi.org/10.5194/acp-14-913-2014

Solberg S, Hov Ø, Søvde A, Isaksen ISA, Coddeville P, de Backer H, Forster C, Orsolini Y, Uhse K (2008) European surface ozone in the extreme summer 2003. J Geophys Res Atmos 113:D07307

Sörgel M, Regelin E, Bozem H, Diesch J-M, Drewnick F, Fischer H, Harder H, Held A, Hosaynali-Beygi Z, Martinez M et al (2011) Quantification of the unknown HONO daytime source and its relation to NO2. Atmos Chem Phys 11(20):10433–10447

SPARC-CCMVal, SPARC Report on the Evaluation of Chemistry-Climate Models (2010) In: Eyring V, Shepherd TG, Waugh DW (eds) SPARC Report No. 5, WCRP-132, WMO/TDNo.1526. http://www.sparc-climate.org/publications/sparc-reports/sparc-report-no5. Last access: Oct. 2012

Steiner AL, Tonse S, Cohen RC, Goldstein AH, Harley RA (2006) Influence of future climate and emissions on regionalair quality in California. J Geophys Res-Atmos 111:d18303

Stevenson DS, Dentener FJ, Schultz MG, Ellingsen K, van Noije TPC, Wild O, Zeng G, Amann M, Atherton CS, Bell N, Bergmann DJ, Bey I, Butler T, Cofala J, Collins WJ, Derwent RG, Doherty RM, Drevet J, Eskes HJ, Fiore AM, Gauss M, Hauglustaine DA, Horowitz LW, Isaksen ISA, Krol MC, Lamarque JF, Lawrence MG, Montanaro V, Muller JF, Pitari G, Prather MJ, Pyle JA, Rast S, Rodriguez JM, Sanderson MG, Savage NH, Shindell DT, Strahan SE, Sudo K, Szopa S (2006) Multimodel ensemble simulations of present-day andnear-future tropospheric ozone. J Geophys Res-Atmos 111:D08301

Stolarski RS, Douglass AR, Oman LD, Waugh DW (2015) Impact of future nitrous oxide and carbon dioxide emissions on the stratospheric ozone layer. Environ Res Lett 10:034011

Stone D, Whalley LK, Heard DE (2012) Tropospheric OH and HO_2 radicals: field measurements and model comparisons. Chem Soc Rev 41:6348–6404

Strauss J, Schirrmeister L, Grosse G, Wetterich S, Ulrich M, Hubberten HW (2013) The deep permafrost carbon pool of the yedoma region in Siberia and Alaska. Geophys Res Lett 40(23):6165–6170

Tagaris E, Manomaiphiboon K, Liao K-J, Leung LR, Woo J-H, He S, Amar P, Russell AG (2007) Impacts of global climate change and emissions on regional ozone and fine particulate matter concentrations over the United States. J Geophys Res-Atmos 112:D14312

Tai APK, Mickley LJ, Heald CL, Wu SL (2013) Effect of CO2 inhibition on biogenic isoprene emission: Implications for air quality under 2000 to 2050 changes in climate, vegetation, and land use. Geophys Res Lett 40:3479–3483. https://doi.org/10.1002/Grl.50650

Taipale R, Kajos MK, Patokoski J, Rantala P, Ruuskanen TM, Rinne J (2011) Role of de novo biosynthesis inecosystem scale monoterpene emissions from a boreal Scots pine forest. Biogeosciences 8(8):2247–2255

Tang Q, Hess PG, Brown-Steiner B, Kinnison DE (2013) Tropsphericozone decrease due to the Mount Pinatubo eruption:Reduced stratospheric influx. Geophys Res Lett 40:5553–5558

Tiwari S, Rai R, Agrawal M (2008) Annual and seasonal variations in tropospheric ozone concentrations around Varanasi. Int J Remote Sens 9(15):4499–4514

Wang W, Tian W, Dhomse S, Xie F, Shu J, Austin J (2014) Stratospheric ozone depletion from future nitrous oxide increases. Atmos Chem Phys 14:12967–12982

Wang S-Y, Schmidtd J, Baidar S, Coburn S, Dix B, Koenig T, Apel E, Bowdalo D, Campos T, Eloranta E, Evans M, DiGangii J, Zondlo M, Gao R-S, Haggerty J, Hall S, Hornbrook R, Jacob D, Morley B, Pierce B, Reeves M, Romashkin P, terSchure A, Volkamer R (2015) Active and widespread halogen chemistry in the tropical and subtropical free troposphere. P Natl Acad Sci USA 112:9281–9286

Weiss M, Bonnel P, Kühlwein J, Provenza A, Lambrecht U, Alessandrini S, Carriero M, Colombo R, Forni F, Lanappe G, Le Lijour P, Manfredi U, Montigny F, Sculati M (2012) Will Euro 6 reduce the NOx emissions of new diesel cars? – Insights from on-road tests with Portable Emissions Measurement Systems (PEMS). Atmos Environ 62:657–665

West JJ, Smith SJ, Silva RA, Naik V, Zhang Y, Adelman Z, Fry MM, Anenberg S, Horowitz LW, Lamarque J-F (2013) Co-benefits of mitigating global greenhouse gas emissions for future air quality and human health. Nat Clim Chang 3:885–889

Wild O (2007) Modelling the global tropospheric ozone budget: exploringthe variability in current models. Atmos Chem Phys 7:2643–2660

Wu SL, Mickley LJ, Jacob DJ, Logan JA, Yantosca RM, Rind D (2007) Why are there large differences between models inglobal budgets of tropospheric ozone? J Geophys Res-Atmos 112:D05302

Wu S, Mickley LJ, Leibensperger EM, Jacob DJ, Rind D, Streets DG (2008) Effects of 2000–2050 global change on ozone air quality in the United States. J Geophys Res 113:D06302

Wu S, Mickley LJ, Kaplan JO, Jacob DJ (2012) Impacts of changes in land use and land cover on atmospheric chemistry and air quality over the 21st century. Atmos Chem Phys 12:1597–1609

Xie X, Shao M, Liu Y, Lu S, Chang CC, Chen ZM (2008) Atmos Environ 42:6000–6010

Xue LK, Wang T, Zhang JM, Zhang XC, Deliger, Poon CN, Ding AJ, Zhou XH, Wu WS, Tang J, Zhang QZ, Wang WX (2011) Source of surface ozone and reactive nitrogenspeciation at Mount Waliguan in western China: new insights from the 2006 summer study. J Geophys Res 116:D07306

Young PJ, Archibald AT, Bowman KW, Lamarque J-F, Naik V, Stevenson DS, Tilmes S, Voulgarakis A, Wild O, Bergmann D, Cameron-Smith P, Cionni I, Collins WJ, Dalsøren SB, Doherty RM, Eyring V, Faluvegi G, Horowitz LW, Josse B, Lee YH, MacKenzie IA, Nagashima T, Plummer DA, Righi M, Rumbold ST, Skeie RB, Shindell DT, Strode SA, Sudo K, Szopa S, Zeng G (2013) Preindustrialto end 21st century projections of tropospheric ozonefrom the Atmospheric Chemistry and Climate Model Intercomparison Project (ACCMIP). Atmos Chem Phys 13:2063–2090

Zeng G, Pyle JA (2003) Changes in tropospheric ozone between 2000 and 2100 modeled in a chemistry-climate model. Geophys Res Lett 30:1392. https://doi.org/10.1029/2002GL016708

Zeng G, Pyle JA, Young PJ (2008) Impact of climate change on tropospheric ozone and its global budgets. Atmos Chem Phys 8(2):369–387

Zeng G, Morgenstern O, Braesicke P, Pyle JA (2010) Impact of stratospheric ozone recovery on tropospheric ozone and its budget. Geophys Res Lett 37:L09805

Zhao Y, Zhang J, Nielsen CP (2013) The effects of recent control policies on trends in emissions of anthropogenic atmospheric pollutants and CO2 in China. Atmos Chem Phys 13:487–508

Zheng X, Shen C, Wan G, Liu K, Tang J, Xu X (2011) 10Be/7Be implies the contribution of stratosphere –tropospheretransport to the winter-spring surface O3 variation observed on the Tibetan Plateau. Chin Sci Bull 56:84–88

Zhou W, Cohan DS, Henderson BH (2014) Slower ozone production in Houston, Texas following emission reductions: evidence from Texas Air Quality Studies in 2000 and 2006. Atmos Chem Phys 14:2777–2788

Chapter 3
Effect of Ozone on Physiological and Biochemical Processes of Plants

Check for updates

Abstract The oxidizing nature of O_3 is responsible for its phytotoxic effects on plants. O_3 enters the plants through stomata and dissolves in the aqueous phase of the substomatal cavity to generate reactive oxygen species (ROS) such as superoxide anions ($O_2^{\circ-}$), hydrogen peroxide (H_2O_2), hydroxyl radicals (OH°) and singlet oxygen ($^1O_2$). Although ROS are an inevitable part of normal cellular metabolism and are continuously produced in the subcellular compartments like mitochondria, peroxisomes, chloroplasts, etc., O_3 exposure stimulates the overproduction of ROS which exceeds the scavenging capacity of the cells intrinsic defense machinery. Plants have incorporated a constitutive antioxidative system which operates to scavenge the ROS generated under normal as well as stress conditions. The defense mechanism of plants has both enzymatic as well as non enzymatic components and works towards annihilating the ROS generated in apoplast as well as symplast. Excess of ROS that are not scavenged by apoplastic antioxidants, target the membrane permeability via the lipid peroxidation of the bilipid layer of the membranes. O_3 also brings about alterations in the physiological process by affecting the biochemistry of photosynthetic machinery, disrupting the chlorophyll fluorescence kinetics and light as well as dark reactions of photosynthesis. O_3 also alters the biophysical parameters like stomatal conductance and internal CO_2 concentration which directly affects the rate of photosynthesis. In addition to this, excess of ROS stimulates the enhanced biosynthesis of cellular antioxidants. O_3 stress also brings about changes in allocation of photosynthates, as more biomass is utilized in O_3 injury repair rather than being converted to storage sugar, starch. O_3 also affects the enzymes of nitrogen metabolism, thus influencing the biosynthesis of amino acids. This chapter investigates the role O_3 in ROS generation and stimulation of antioxidant production. Effect of O_3 on physiological processes, metabolite contents and nitrogen metabolism is also discussed.

Keywords ROS · Scavenging capacity · Apoplastic antioxidants · Lipid peroxidation

Contents

© Springer International Publishing AG 2018
S. Tiwari, M. Agrawal, *Tropospheric Ozone and its Impacts on Crop Plants*,
https://doi.org/10.1007/978-3-319-71873-6_3

1   Introduction

The phytotoxic nature of O_3 can be attributed to its ability to act as a powerful oxidizing agent, leading to the generation of additional Reactive Oxygen Species (ROS) which disrupts plant metabolism. ROS generation is an unfavorable consequence of aerobic metabolism (Miller et al. 2010; Temple et al. 2005). It has been estimated that 1–2% of O_2 consumption is used up in the formation of ROS in tissues (Bhattacharjee 2005). Under normal metabolic activity, ROS are continuously produced in sub-cellular compartments like chloroplasts, mitochondria, peroxisomes and a few other sites in the cell due to their O_2 producing metabolic activities such as photosynthesis and respiration (Tripathy and Oelmüller 2012). The commonly produced ROS are $^1O_2$, $O_2^{\circ-}$, H_2O_2 and OH°. The oxidizing nature of ROS may pose a threat to the normal cellular functioning, and if uncontrolled can lead to cell death (Miller et al. 2010). However, beside acting as harmful oxidizing agents, ROS play a significant role in fundamental plant processes such as stomatal closure, response to environmental stresses, modulation of protein activity, gene expression, etc. (Moller et al. 2007; Apel and Hirt 2004). At low concentrations, ROS also act as important signaling molecules (Tripathy and Oelmüller 2012). ROS damage the nuclear material of the cell and has harmful effects on mitochondrial, chloroplastic and cytosolic DNA as well. Excessive ROS generation results in deoxyribose strand breakage, removal of nucleotides, modification in organic bases of nucleotides and DNA-protein crosslinks, thus increasing the frequency of mutation (Petrov and Breusegem 2012; Sharma et al. 2012; Imlay and Linn 1998). Mitochondrial and chloroplastic DNA are more sensitive to oxidative damage than nuclear DNA due to lack of protective histone proteins and because of their close proximity to ROS generating centres (Shokolenko et al. 2014; Richter 1992). The role of ROS as a damaging entity or as a signaling molecule, depends upon the site or time of their production (Schieber and Chandel 2015). At low concentrations, ROS act as signals of defense pathway in eliciting cellular responses (Dietz 2016; Baier et al. 2005). However, at high concentrations, ROS can cause cell death, either by ROS toxicity or through specific ROS activated signaling cascade leading to programmed cell death (Karuppanapadian et al. 2011; Quan et al. 2008).

2 Ozone Uptake

The O_3 flux based approach has been recommended as a suitable indicator of O_3 damage in Europe (UNECE 2004). Experimental evidences indicate that there exists a strong relationship between O_3 uptake and stomatal conductance in trees such as *Quercus ilex, Q. robur, Q. pubescens* and *Populus nigra* (Loreto and Fares 2007). Beside this, the uptake of O_3 in darkened leaves in which stomata were allowed to close was less than 10% of that recorded in illuminated leaves (Altmiret et al. 2006). It has been established that impact of O_3 on plants not only depends upon the concentration of O_3 but on the rate of O_3 uptake by plants (Mills et al. 2011). The flux of O_3 from the atmosphere into the plant depends upon different resistances acting at various levels i.e. aerodynamic resistance depending upon the atmospheric turbulence, boundary layer resistance due to the air adjacent to the leaf, stomatal resistance caused by the stomatal pores and an internal resistance in the mesophyll cells of the plant leaf (Fig. 3.1).

O_3 uptake through stomata is also regulated by establishment of continuous gradient between the external O_3 concentration and concentration of O_3 in the substomatal cavity which is assumed to be zero (Loreto and Fares 2007). In case of high stomatal conductance, there occurs a strong inflow of O_3 inside the leaf which saturates the scavenging potential of antioxidative defense system of plants (Caregnato et al. 2013). As such, O_3 accumulates in the substomatal cavity, thereby reducing the gradient with external O_3 concentration. For calculating the uptake of O_3 through stomata, not only the concentration of O_3 at the plant height and stomatal resistance (R_s) is important, but the leaf boundary layer resistance (R_b) also plays an important role. The uptake of O_3 (O_3 uptake) in the plants is expressed by the formulae given by Danielsson et al. (2013).

However, O_3 flux is not always a suitable damage index as it excludes the detoxification component of plants (Gerosa et al. 2013). Once the stomatal flux is known, it is possible to calculate the O_3 dose, which is simply the sum of stomatal O_3 fluxes (Mills et al. 2011). UN/ECE scientific community has introduced the term

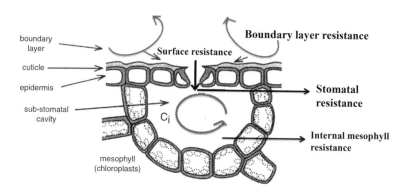

Fig. 3.1 Different types of resistances determining the ozone flux in the leaves

"Phytotoxical Ozone Dose" (POD_1), which is the cumulative O_3 dose over a threshold of 1 nmol O_3 m^{-2} s^{-1} (Mills et al. 2011). POD_1 takes into account the internal potential of the plants to detoxify part of O_3 entering through stomata. POD_1 is shown to be better correlated with the biomass reduction than simple O_3 dose, thereby facilitating the evaluation of harmful effects of O_3 on plants (Gerosa et al. 2013).

3 ROS Generation in the Cell

Biotic stresses like pathogen and insect attack and abiotic stresss like heavy metal, drought, salinity, low temperature, pollution, etc. lead to the overproduction of deleterious chemical entities called reactive oxygen species (ROS) (Sharma et al. 2012; Tripathy and Oelmüller 2012). The relative contents of different ROS depend upon the kind of stress to which the plant is exposed. For instance, high light intensity leads to the formation of $_1O^2$ which is primarily formed by energy transfer from triple chlorophyll to triplet O_2 (Tripathy and Oelmüller 2012). As an important byproduct of aerobic respiration, ROS are produced in cellular compartments like mitochondria, chloroplasts, peroxisomes etc. which involve a constant flow of electrons (Dietz et al. 2016; Gilory et al. 2016; Apel and Hirt 2004). Within the cell, ROS are produced mostly in chloroplasts and peroxisomes, while mitochondria have a less significant contribution (Foyer and Noctor 2000). In addition to these cell organelles, apoplastic space in plants also acts as an important site of ROS generation (Vaahtera et al. 2014; Gilroy et al. 2016; Mignolet- Sprut et al. 2016). The inhibition of electron transport under stress conditions leads to the over reduction of many component of electron transport chain (Roach and Krieger-Liszkay 2014). ROS can also be generated through reduced activity of Rubisco, under abiotic stress conditions (Ashraf and Harris 2013). Abiotic stress generally reduces CO_2 availability due to stomatal closure and enhances the production of ROS such as $O_2^{\circ-}$ and $_1O^2$ in chloroplasts (Mignolet-Spruyt et al. 2016). Chlorophyll acts as the main light absorbing pigment and is present in both light harvesting complexes (LHC) and photosynthetic reaction centres. Normally, the electron flow from the excited photosystem centers is directed to $NADP^+$, which is reduced to NADPH. It then enters the Calvin cycle and reduces the final electron acceptor, CO_2. In case of overloading of electron transport chain (ETC), ROS are generated, which disturb the photosystem stoichiometry and lead to the diversion of a few electrons from ferridoxin to O_2, reducing it to $O_2^{\circ-}$ via Mehler's reaction (Pesaresi et al. 2009; Elstner 1991). The $O_2^{\circ-}$ formed is dismutated to H_2O_2 by CuZn SOD on the stromatal membrane surface (Takahashi and Asada 1988). Chloroplasts are the main production site of $^1O_2$ which is produced through energy transfer from triplet chlorophyll (Chl*) to molecular O_2 (Roach and Krieger- Liszkay 2014).

Peroxisomes are also important sites of H_2O_2 production due to their essentially oxidative type of metabolism (Tripathy and Oelmüller 2012; Del Rio et al. 2006). Two important sites recognized for $O_2^{\circ-}$ production are located in the peroxisomal matrix, where xanthane oxidae catalyses the oxidation of xanthine and hypoxanthine to uric acid and the peroxisomal membrane, where an electron transport chain operates and produces $O_2^{\circ-}$ (Del Rio et al. 2006). H_2O_2 is produced in the peroxisomes through oxidative processes like photorespiration and fatty acid β-oxidation, occurring in the matrix (Baishanab and Ralf 2012; Kerchev et al. 2016). In mitochondria, ROS are generally formed in NADH dehydrogenase (Complex I) and ubiquinone- cytochrome region (Complex III) during electron transfer through electron transport chain (Quinlan et al. 2013). In plants, under normal aerobic conditions, electron transport chain and ATP synthesis are tightly coupled. However, various stress factors cause the uncoupling of the two processes, leading to over reduction of electron carriers and hence ROS formation (Blokhina and Fagerstedt 2010). In mitochondria, $O_2^{\circ-}$ is the primary ROS produced, which is quickly converted to membrane permeable H_2O_2 or extremely active OH° radicals (Huang et al. 2016). When NAD^+ linked substrate for complex I become limited, electron transport can occur from complex II to complex I (reverse electron flow), which increases ROS production at complex I (Huang et al. 2016).

In endoplasmic reticulum, NAD(P)H dependant electron transport involving Cyt P450 produces $O_2^{\circ-}$ (Mittler 2002). In plasma membrane, two enzymes NADPH oxidase and quinine reductase are responsible for production of $O_2^{\circ-}$. NADPH oxidase contains a multimeric flavocytochrome that forms ETC capable of reducing O_2 to $O_2^{\circ-}$, whose production is impaired by inhibitors of this oxidase (Sharma et al. 2012). In addition to this oxalate oxidase and amine oxidase associated with cell wall, also generate H_2O_2 in apoplast (Hu et al. 2008). Cell wall associated peroxidases also act as important sources of H_2O_2 (Sharma et al. 2012; Gross 1977). Figure 3.2 summarizes the different sites of ROS production in the cell.

O_3 stress elicits a series of events in the plant cells that are similar to the pathogen defense responses (Sandermann et al. 1998). It enters the leaves through stomata and reacts with the aqueous phase of the apoplast forming ROS, specially $O_2^{\circ-}$ and H_2O_2 (Ernst et al. 2012). Apoplastic ROS generation mechanism plays a significant role in producing oxidative stress in plants exposed to high O_3 concentration. Jajic et al. (2015) suggested H_2O_2 to be the main damaging chemical entity produced in the apoplast, in addition to other ROS (Tripathy and Oelmüller 2012). A biphasic oxidative burst involving NADPH-dependant oxidases (Rao and Davis 1999), resulting in generation and accumulation of H_2O_2 and $O_2^{\circ-}$, has been proposed to be the main mechanism of O_3 induced cell death (Langsbartels et al. 2002). NDPH oxidase is an enzyme complex found associated with plasma membrane, utilizing NADPH as electron donar. This complex oxidizes NADPH at the cytosolic surface and reduces O_2 to $O_2^{\circ-}$ at the outer surface (Kaur et al. 2014; Sagi and Fluhr 2006). H_2O_2 in the apoplast is produced indirectly by spontaneous or SOD mediated dismutation (Gilroy et al. 2016). In plants, NADPH oxidases are known as respiratory burst oxidase homologs (Rbohs) (Marino et al. 2012; Sagi and Fluhr 2006). Rbohs play an active role in generation of ROS not only in the plasma membrane, but also in the membranes surrounding the cellular organelles. Joo et al. (2005) showed that ROS production was initiated from guard cell chloroplasts, followed by ROS production in guard cell membranes which required NADPH oxidase encoded by

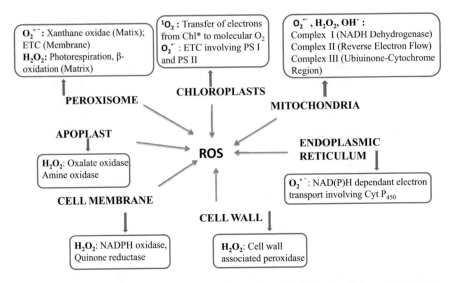

Fig. 3.2 Different types of reactive oxygen species produced in different cellular compartments

AtrbohD and AtrbohF genes. Kangasjarvi et al. (2005) also suggested NADPH oxidase to be an important source of ROS production under O_3 stress. However, Ueda et al. (2013) did not found any variation in the activity of NADPH oxidase in rice plants (*Oryza sativa* L. var. Koshihikari) grown under 6 h O_3 concentration of 150 ppb. In *Arabidopsis* Col-0 treated with 250 ppb O_3, the first phase of O3 induced ROS accumulation was significantly reduced, but not abolished in AtrbohD and AtrbohD/F mutants, which suggest that there may be several other sources (like cell wall peroxidases) of ROS production, in addition to NADPH oxidases (Vahisalu et al. 2010). These results indicate that more experiments are needed to clearly understand the ROS generation in plants under O_3 stress and the role of NADPH oxidase in this process.

Reduction in the PS II efficiency is a well known feature of plants exposed to O_3 stress. Feng et al. (2010) and Singh et al. (2009) recorded significant reductions in PS II efficiency in wheat and soybean cultivars, respectively which indicates a disruptive electron transport chain leading to the generation of $O_2{}^{\circ-}$. O_3 treated leaves observed a higher number of deactivated PS II centres and decreased photochemical activity, which caused leakage of electrons resulting in increased generation of ROS (Biswas et al. 2013). Feng et al. (2016) observed slower electron transport rates in five modern cultivars of Chinese winter wheat varieties (*Triticum aestivum* L) grown at elevated O_3 concentrations. Ueda et al. (2013) recorded significant level of down-regulation in the gene expression of chloroplastic SOD which can be attributed to the enhanced ROS ($O_2{}^{\circ-}$ and H_2O_2) in chloroplasts under O_3 fumigation. In addition to these, dark respiration in mitochondria and photorespiration in peroxisomes are up-regulated under acute O_3 stress, which would lead to enhanced ROS generation in mitochondria and peroxisomes (Kangasjarvi et al. 2005).

Under steady state condition, ROS generated as a byproduct of different metabolic pathways are scavenged by various enzymatic and non enzymatic antioxidants
(Wang et al. 2014; Rai and Agrawal 2014). Under stress conditions, ROS production within the cell increases abruptly, a condition called "oxidative burst" disrupting the cellular homeostasis (Mullineaux and Baker 2010; Sharma et al. 2012).
Increased levels of intrinsic ROS produced under stress conditions may exceed the
antioxidative defense capacity of the plants resulting in significant damage to cellular structure (Foyer and Noctor 2005). It is important to note, whether ROS will
act as a damaging, protective or signaling molecule, as it depends upon the equilibrium between ROS production and scavenging at proper site and time (Gratao et al.
2005).

4   Defense Strategy of Plants

ROS generation in plant under O_3 stress has led to the evolution of specific metabolic pathways, dedicated to protect the plants from ROS toxicity (Foyer and
Noctor 2013). If unchecked, ROS may get accumulated in the cell resulting in
oxidative damage to membranes (lipid peroxidation), oxidation of sulphur containing residues of proteins and alteration of protein structure and formation of disul-
phide bond, oxidation of nucleic acid, etc. (Mittler 2002; Foyer and Noctor 2013;
Dietz et al. 2016). Plants have already incorporated a constitutive detoxification
system comprising of enzymatic and non enzymatic antioxidants, which operates
at the entrance of O_3 into the leaf (apoplast), as well as in different sub cellular
compartments (symplast). The apoplastic and the symplastic antioxidants work
together to keep ROS production under control with each antioxidant specifically
scavenging a particular ROS (Table 3.1). The apoplastic antioxidants, specially the
pool of reduced ascorbate (AsA) scavenge O_3 and O_3 generated ROS thus serving
as first detoxification barrier of the cell (Fig. 3.3) (Heath 2008). Other important
antioxidants reported from apoplastic spaces are SOD and APX which together
aim to scavenge $O_2^{\circ-}$ and H_2O_2 formed in the apoplast (Mittler 2002; Gill and
Tuteja 2010). SOD acts as an important plant stress tolerance enzyme helping as
first line of defense by scavenging $O_2^{\circ-}$ by catalyzing the dismutation of two molecules of $O_2^{\circ-}$. It removes $O_2^{\circ-}$ and hence reduces the possibility of formation of
OH radicals by Haber-Wies type reaction (Ueda et al. 2013; Sharma et al. 2012).
Γ-Glutamylcysteine synthatase is a reducing cofactor for several enzymes involved
in ROS detoxification, thus playing an important role in oxidative defense system
of plants (Noctor et al. 2012).

AsA is the most abundant and powerful antioxidant present in most of the cell
compartments including apoplast where it plays a significant role in scavenging the
O_3 generated ROS (Feng et al. 2014; Athar et al. 2008). It has been reported that
ascorbic acid generally remains available in reduced form (ASH) (Booker et al.
2012). The oxidized forms of ascorbic acid (MDHA and DHA) have a short half life
and would be lost unless converted to reduced form. AsA and GSH are two important

Table 3.1 Different antioxidant enzymes of the plant defense machinery and the reactions catalyzed by them

	Antioxidant enzyme	Reaction catalyzed
1.	Superoxide dismutase (SOD)	$O_2^\cdot + O_2^\cdot + 2H^+ = 2H_2O_2 + O_2$
2.	Catalase (CAT)	$H_2O_2 = H_2O + \frac{1}{2}O_2$
3.	Ascorbate peroxidase (APX)	$H_2O_2 + AA = H_2O + DHA$
4.	Guaicol peroxidase (GuPX)	$H_2O_2 + GSH = H_2O + GSSH$
5.	Monohydroascorbate reductase (MDHAR)	$MDHA + NAD(P)H = AA + NAD(P)^+$
6.	Dehydroascorbate reductase (DHAR)	$DHA + 2GSH = AA + GSSH$
7.	Glutathione reductase (GR)	$GSSH + NAD(P)H = 2GSH + NAD(P)^+$

$O_2^\cdot$ =Superoxide radical; H_2O_2 = Hydrogen peroxide; AA = Ascorbic acid; GSH = Reduced gluta-thione; GSSH = Oxidized glutathione

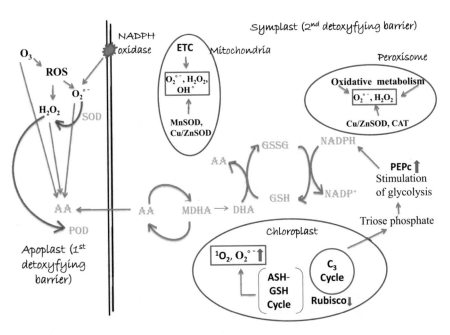

Fig. 3.3 Mechanism of action of antioxidant defense machinery of plants

antioxidants of the plant defense system that constitute the AsA-GSH cycle which forms an important component of plant's antioxidant defense system. Through ASH-GSH pathway, ascorbic acid in its reduced form functions as an electron donor and scavenges the apoplastic ROS acting as a substrate for APX (Gill and Tuteja 2010; Foyer and Noctor 2005). Dumont et al. (2014) observed significant increase in AsA and GSH contents in three genotypes of poplar (*Populus deltoids* Bartr. x *Populus nigra* L.) treated with 120 ppb O_3 (13 h) for 17 days. The level of apoplastic ascorbate is considered to be a good indicator of O_3 tolerance (Dumont et al. 2014; Di Bacchio et al. 2008; Tausz et al. 2007). The high GSH/GSSG and AsA/DHA ratio is considered to be an adaptation in plants in maintaining an appropriate redox environment and reducing the oxidative stress caused by O_3 (Dumont et al. 2014). The main target of APX in the apoplastic region is H_2O_2, thus the enzyme shows high affinity for H_2O_2 (µM range) than CAT and POD (mM range) (Gill and Tuteja 2010). H_2O_2 generated in the apoplast is reduced to water, through a reaction catalyzed by APX. The resulting byproducts MDHA and DHA are directly reduced to ascorbate by NAD(P)H or NADH as electron donors (Foyer and Noctor 2011). However, under O_3 stress, the normal cellular homeostasis is disturbed, leading to the accumulation of H_2O_2 in the apoplast. The accumulated apoplastic H_2O_2 can be transported to different cellular compartments in a regulated process via aquoaporins (Tian et al. 2016).

When ROS production crosses the scavenging capacity of the apoplastic antioxidants, depletion of antioxidants occur, which leads to membrane injury (Feng et al. 2016; Conklin and Barth 2004; Zheng et al. 2000). Membranes are the primary target sites for O_3 injury (Heath and Taylor 1997). Metabolic decomposition of membrane glycolipids and increased amounts of triacylglycerol and fatty acids have been observed in O_3 exposed plants (Sakaki et al. 1990a, b, c). The effect of O_3 on membrane lipid composition leads to membrane leakage (Marre et al. 1998), which disturbs cellular homeostasis (Mullineaux and Baker 2010; Sharma et al. 2012). ROS stimulates the exchange of antioxidants between symplastic detoxification system and apoplastic space (Diengremel et al. 2008). Ascorbate regeneration is tightly coupled to GSH (reduced Glutathione) within the cell and the ascorbic acid thus generated is transported to apoplast to replenish the reduced ascorbate pool (Baier et al. 2005). The capacity of the plants to regenerate the antioxidants within the cells serves as the second detoxification barrier (Fig. 3.3). The regeneration of ascorbate depends upon the ASH-GSH cycle, the reducing power required to accomplish this process is supplied by NAD(P)H (Dizengremel et al. 2008). Beside ascorbate (reduced and oxidized) and APX, the other important antioxidant entity which contributes to the symplastic defense mechanism is CAT. CAT is responsible for the scavenging of H_2O_2, which is a powerful oxidant, rapidly oxidizing thiol groups (Garg and Manchanda 2009). CAT has an important role to play in chloroplasts which contain several thiol regulated enzymes involved in Calvin cycle (Kaiser 1979). In the absence of CAT from chloroplasts, accumulation of H_2O_2 in the organelle occurs, which oxidizes the thiol based enzymes present therein (Kaiser 1979). CAT has a vital role in the removal of H_2O_2 generated in peroxisomes by oxidases involved in β-oxidation of fatty acids, photorespiration and purine catabolism. An alternative mode of H_2O_2 destruction within the symplast is by peroxidases, specifically APX. Another enzyme contributing towards the symplastic defense is glutathione reductase (GR) which catalyses the reduction of GSSH (oxidized glutathione) to GSH (reduced glutathione) (Foyer and Noctor 1998). The GSH/GSSH couple plays a significant role in regeneration of reduced ascorbate (Foyer and Noctor 2005). GSH not only serves as a reducing co- factor for several enzymes involved in ROS detoxification, but also conjugates to specific proteins to prevent their oxidation (Rouhier et al. 2004).

Researches conducted in the last few years have proved that several biogenic volatile organic compounds (BVOCs) play important roles in providing antioxidative defense to plants against O_3 stress (Loreto et al. 2004; Valikova et al. 2012; Possell and Loreto 2013). Several plants posses an intrinsic capability to synthesize stress ethylene in response to O_3 stress (Velikova et al. 2012; Heiden et al. 1999). O_3 sensitive tobacco cultivar (*Nicotiana tabacum* L. Bel W3) showed a significant increase in ethylene synthesis after O_3 exposure, whereas it remains low in insensitive cultivar Bel B (Langbartels et al. 1991). The glandular trichomes found on the leaf surface of *Nicotiana tabaccum* var. Ambalema exude diterpenoid, cis-abienol, which cover the plant leaves as defense barrier and act as a chemical protection shield against stomatal O_3 uptake by depleting O_3 at the leaf surface (Jud et al. 2016). Due to ozonolysis at the leaf surface, Ambalema kept the O_3 conductance at elevated levels for many hours as compared to 3H02 cultivar which does not exude diterpenoids (Jud et al. 2016). Studies have shown that both ethylene evolution and concentration of ethylene precursor ACC increase rapidly after 1–2 h of O_3 exposure (Vahala et al. 1998; Overmeyer et al. 2000; Tamaoki et al. 2003). Along with the formation of stress ethylene, polyamine pathway is also stimulated during O_3 stress (Jang et al. 2012). Polyamines are nitrogenous bases belonging to fatty acid groups with biological activity that are produced during metabolic processes. Upon O_3 exposure, arginine carboxylase, which is an important enzyme of polyamine pathway is induced which results in an increasing level of spermidine in barley (Zhao et al. 2007) and putrescine in O_3 resistant tobacco (van Buuren et al. 2002). Polyamines increase the stress tolerance in plants by preventing O_3 induced lipid peroxidation by chelating metal ions that catalyse this reaction (Toderova et al. 2015). O_3 stress has been shown to increase the activities of enzymes such as PAL (phenylalanine ammonium lyase) in *Cicer arietinum* and *Trigonella foenum* (Srinandhinidevi et al. 2015), CHS (chalcone synthase) in *P. vulgaris* (Paolacci et al. 2001) and CAD (cinnamyl alcohol dehydrogenase) in *Pinus sylvestris*, which control the phenylpropanoid, flavinoid and lignin pathways, respectively.

5 Membrane Permeability

It is well cited that O_3 can disrupt the membrane permeability by disorganizing the membrane structure (Calatayud et al. 2003; Tripathy and Agrawal 2013). O_3 primarily targets the lipid component of the membrane leading to its peroxidation (Whittaker et al. 1990; Foyer and Noctor 2005; Sharma et al. 2012). O_3 induced lipid peroxidation may increase production of ROS including $O_2^{\circ-}$ and H_2O_2 (Mishra and Agrawal 2015). With respect to photosynthetic apparatus of plants growing under O_3 stress, dysfunctioning of membrane permeability in chloroplast has an adverse effect upon membrane bound reaction centres in electron transport chain, inactivating the rate of photosynthesis and activating respiration (Foyer and Noctor 2011). Increments in lipid peroxidation upon increased ROS production have been reported in rice (Rai and Agrawal 2008), wheat (Mishra et al. 2013), radish and brinjal (Tiwari and Agrawal 2011), carrot (Tiwari et al. 2006), clover (Calatayud et al. 2002), lettuce (Francini et al. 2007) and pea (Carlsson et al. 1996). It has been shown that O_3 stress stimulated peroxidation of membrane lipids in lentils, increasing its hydroperoxide content especially in esterified fatty acid (Maccarrone et al. 1997). O_3 is known to modulate the expression of lipoxygenase which targets the lipid ester fraction of biological membranes and plays an important role in determining the O_3 sensitivity of plants (Francini et al. 2007).

Lipid peroxidation is generally measured as malonaldehyde (MDA) content and has been correlated with the degree of injury to the membrane under O_3 exposure (Ranieri et al. 1996). MDA is the end product of ROS induced oxidation of polyunsaturated fatty acids; therefore it is frequently used as a biomarker of oxidative stress (Bhatacharjee 2015). Excess ROS production under O_3 stress hinders the natural potential of specific cell wall components to scavenge the overproduced ROS, leading to damage to the plasma membrane (Sammartin et al. 2003). According to Heath (2008) the peroxidation of lipid components results in solute leakage due to altered membrane permeability. Liu et al. (2015) studied the response of wheat to elevated O_3 and observed that MDA content increased by 314.3% and 65% at jointing and heading stages, when treated with 120 ppb (4 hd^{-1}) O_3. Mishra and Agrawal (2015) correlated lipid peroxidation and solute leakage in *Vigna radiate* L.cv HUM-2 and HUM-6 grown at 8 h mean O_3 concentration of 68.9 ppb and found that both the parameters showed similar incremental trends, the magnitude of response being higher in sensitive cultivar HUM-2 than the tolerant cultivar HUM-6. Multivariate ANOVA test showed that lipid peroxidation in both the cultivars showed significant variations due to O_3 treatments (Mishra and Agrawal 2015). Lipid peroxidation and solute leakage increased significantly by 30.8% and 42%, respectively in HUM-2 and by 21% and 26.5%, respectively in HUM-6 (Mishra and Agrawal 2015). Similar increasing trends in lipid peroxidation and solute leakage were also observed in *Citrus clementina* Hort (ex. Tan) cv. Marisol grown at mean O_3 concentration of 30–65 ppb (Iglesias et al. 2006). Exogenous application of ascorbic acid, an antioxidant reduced the degree of lipid peroxidation by regeneration of reduced tocopherol which favours the formation of lipid hydrophilicity thus controlling lipid peroxidation of membrane lipids (Blokhina et al. 2003). This was evident by the reduction in degree of lipid peroxidation in *V. radiate* L. cv. HUM-1 (13.7% and 13%) and cv HUM-24 (8.9% and 7.3%) exposed to 12 h mean O_3 concentrations of 65.3 and 71.1 ppb and sprayed with 12.5 mM freshly prepared ascorbic acid solution (Chaudhary and Agrawal).

Sensitive wheat cultivars *T.aestivem* L. cv M234 showed higher increments in MDA content (47.4%) at mean O_3 concentrtion of 41 ppb (Rai et al. 2007) as compared to the increments in comparatively resistant cultivars PBW 343 (34.6%) and M533 (4.5%) grown at mean O_3 concentration of 53 ppb (Rai and Agrawal 2014). Radish (*Raphnus sativus* L cv Pusa Reshmi) and brinjal (*Solanum melanogena* L. Pusa hybrid-6) grown at 8 h mean O_3 concentration of 4 0.8 ppb showed significant increments in MDA content, the degree of lipid peroxidation being higher in case of radish than brinjal suggesting that radish was more susceptible to O_3 injury than brinjal (Tiwari and Agrawal 2011). Sarkar et al. (2010) experimenting on two high yielding wheat cultivars observed that lipid peroxidation increased by 24% and

41.4% in cv Sonalika and by 44.4% and 51.1% in HUW510 at mean O_3 concentrations of 50.4 and 54.9 ppb, respectively.

6   Effect of Ozone on Physiological Processes

The negative effects of ozone on physiological processes including reductions in stomatal conductance, net photosynthetic CO_2 assimilation and carboxylation efficiency are well cited (Tetteh et al. 2015; Zhang et al. 2014; Rai and Agrawal 2014; Chaudhary and Agrawal 2014; Singh et al. 2009; Wittig et al. 2009). O_3 at high concentrations negatively affect crops causing physiological modifications, which are translated into morphological changes such as growth (Zhang et al. 2014; Biswas et al. 2008) and yield reductions (Teeth et al. 2015; Wilkinson et al. 2011). Morgan et al. (2003) showed impairment of photosynthetic carbon acquisition in soybean at chronic O_3 exposure of 70 ppb through meta analysis based on 53 peer reviewed studies. O_3 induced loss in photosynthetic rate was attributed to impaired activity of mesophyll cells and loss of integrity of cellular membranes as evident by increased intercellular CO_2 concentration and lipid peroxidation (Biswas et al. 2008). Reductions in physiological parameters like rate of photosynthesis (*Ps*) and stomatal conductance (*gs*) in several crops under variable O_3 concentrations has been widely reported (Yendrek and Ainsworth 2015; Tetteh et al. 2015; Niu et al. 2014;Mishra and Agrawal 2015; Singh et al. 2009).

Ps is a multistep process and its response to O_3 stress depends upon the fact that which particular event of the entire process is affected. Alteration in *Ps* may be influenced either by changes in internal CO_2 or by variations in the light reactions (light energy utilization and conversion) or dark reaction (carboxylation efficiency of RuBisCO). Rai and Agrawal (2008) reported a significant reduction in photosynthetic rate in two rice cultivars, NDR 97 (18%) and Saurabh 950 (28.3%) grown under 8 h mean O_3 concentration of 35 ppb. Meta analysis of 53 peer reviewed studies published between 1980 and 2007 showed that different varieties of wheat when grown in O_3 concentrations with a range of 31–200 ppb showed a reduction of 20% in *Ps*, 22% in *gs*, 19% in RuBisCO activity and 40% in chlorophyll content (Feng et al. 2008). Feng et al. (2011) experimenting on two cultivars of winter wheat Yanmai 16 and Yangfumai 2 exposed to 127% higher than ambient O_3 reported more reductions in *Ps* of O_3 sensitive Yangfumai 2 cultivar. Higher reductions in *Ps* of O_3 sensitive cultivar Sonalika as compared to less sensitive cultivar HUW234 of wheat was reported under O_3 stress (Sarkar et al. 2010). However, contrasting results were obtained by Singh et al. (2009) in soybean grown at 70 ppb O_3, where O_3 sensitive cultivar PK472 showed lesser reduction in *Ps* (19.83%) as compared to O_3 tolerant cultivar Bragg (25.60%). A linear relationship between *Ps* reductions and O_3 levels was recorded by Sun et al. (2014) for two soybean cultivars Dwight and IA3010 grown under O_3 levels ranging between 37–110 ppb. Similar correlation was also observed between reduced growth and reduction in *Ps* in *Trifolium repens* and *Lolium perenne* grown under O_3 concentration 30 ppb above ambient concentration (30+ AA) (Hayes et al. 2009).

6.1 Biochemistry of Photosynthetic Machinery

Exposure to high O_3 concentrations damages the photosynthetic machinery of the plants causing a significant reduction in carboxylation efficiency, which plays a major role in impairment of the process of photosynthesis (Guidi and Degl'Innocenti 2008; Calatayud et al. 2003). O_3 can alter light reaction, decreasing the electron transfer between the two photosystems (Pellegrini et al. 2011; Calatayud et al. 2003). The Fv/Fm ratio is reported to be a good indicator of photoinhibitory damage (Bolhar-Nordekampf et al. 1989). Chlorophyll fluorescence techniques have been used to follow primary processes in photosynthesis (Murchie and Lawson 2013). Guidi et al. (1997) working on two cultivars of *Phaseolus vulgaris* exposed to O_3 concentration of 80 ppb (single 4 h period pulse) reported that Fv/Fm ratio, which indicates the efficiency of excitation capture of PS II, decreased significantly suggesting that electron transport around PS II has been altered.

The chlorophyll fluorescence kinetics also referred to as Kautsky transient, as explained by Krause and Wies (Krause and Wies 1991) upon saturated illumination reveals detailed information not only on the redox state of primary acceptor Q, but also on the electron transport beyond Q_A on the donor side of PS II. The different levels observed in Kautsky transient (Fo and Fm) are used to calculate the electron flow around PS II and Fv/Fm value gives the maximum quantum yield of PS II (Kitajima and Butler 1975). Decrease in Fm corresponds to a transient reduction in the pool of primary PS II electron receptor, Q_A, whereas increase in Fm suggest that reaction centres in electron transport chain remain closed for most of the time and the quantum energy absorbed was mostly dissipated as heat (Singh et al. 2009). O_3 treated leaves also showed a significant increase in Fo which is an indicative of a higher number of deactivated PS II centres and decrease in photochemical efficiency (Guidi and Degl'Innocenti 2008). *Liriodendron tulipifera* grown at O_3 concentration of 120 ppb (5hd^{-1} for 45 days) showed a reduction of 8% in Fv/Fm ratio which was caused due to an increase in Fo and a reduction in Fm (Pellegrini et al. 2011), which is a possible sign of inactivation of PS II centres (Reichenauer et al. 1997). In many leaves of healthy plants, Fv/Fm ratio is generally around 0.8, a value generally accepted for non stressed conditions. However, a reduction in Fv/Fm ratio indicates photoinhibition of photosynthesis (Baker and Rosenqvist 2004). Reduction in the value of Fv/Fm ratio has been reported in wheat (Sarkar et al. 2010; Rai and Agrawal 2014), rice (Sarkar et al. 2015), soybean (Singh et al. 2009; Betzelberger et al. 2012), mung bean (Mishra and Agrawal 2015), marigold and rose (Yang et al. 2016), etc. (Table 3.2).

The disruption of light reaction of photosynthesis upon O_3 stress can also be attributed to the vulnerability of D1 core protein of PS II (Adir et al. 2003). Godde and Buchhold (1992) have shown that O_3 exposure stimulates both synthesis and

Table 3.2 Effect of Ozone on different physiological parameters of selected crop plants

Plants/Cultivars	O₃ concentration (ppb)	Percentage change (%)				Reference
		Photosynthetic rate (P_s)	Stomatal conductance (g_s)	Internal CO₂ (C_i)	Fv/Fm	
Trifolium repens L. NC-S	200 ppb (single pulse for 5 h)	(−) 56.7	NS	(+) 26.2	(−) 4.14	Francini et al. (2007)
Trifolium repens L. NC-R		(−) 26.4	(−) 10	(+) 13.3	(−) 13.79	
Trifolium alexandrium L. (6 cultivars)	57.5 ppb (8 h mean)	(−) 12.7–31.5	(−) 5–29.4	–	(−) 16.6–19.2	Chaudhary and Agrawal (2013)
Spinacia oleracea L.	2400 (AOT40)	(−) 25	(−) 6	(+) 10.56	NS	Calatayud et al. (2003)
	15,048 (AOT40)	(−) 63	(−) 46	(+) 28.4	NS	
Raphnus sativa L. var Pusa reshmi	40.8	(−) 18.82	(−) 54.62	(−) 20.8	(−) 6.9	Tiwari and Agrawal (2011)
Solanum melanogena L. var Pusa hybrid-6		(−) 31.9	(−) 14.74	(−) 14.74	(−) 2.61	
Dacus carota L. Pusa Kesar	30.37	(−) 12.06	(−) 17.25	NS	(−) 5.70	Tiwari et al. (2006)
Glycine max L. PK472	70 ppb (4 hd⁻¹)	(−) 19.83	(−) 21	(+) 13.12	(−) 16.02	Singh et al. (2009)
	100 ppb (4 hd⁻¹)	(−) 40.40	(−) 26.3	(+) 40.87	(−) 6.96	
Glycine max L. Bragg	70 ppb (4 hd⁻¹)	(−) 25.60	(−) 61.35	(+) 1.02	(−) 2.5	Singh et al. (2009)
	100 ppb (4 hd⁻¹)	(−) 32.35	(−) 66.10	(−) 8.53	(−) 5.8	
Glycine max (10 cultivars)	46 ppb	(−) 11	(−) 15	–		Betzelberger et al. (2010)

Vigna radiate L.	cv HUM-1	64 ppb (12 h mean)	(−) 26.3	(−) 13.69	(−) 9.7	(−) 16.66	Chaudhary et al. (2013)
	cv HUM-2		(−) 23.30	(−) 4.6	(−) 8.9	(−) 8.47	
	cv HUM-6		(−) 0.86	(−) 3.15	(−) 8.6	(−) 6.25	
	cv HUM-23		(−) 15	(−) 2.5	(−) 8.2	(−) 4.93	
	cv HUM-24		(−) 19.5	(−) 0.6	(−) 8.67	(−) 5.26	
	cv HUM-26		(−) 18.54	(−) 1.20	(−) 9.65	(−) 5.40	
	cv HUM- 2	68.9 ppb (8 h mean)	(−) 28.3	(−) 38.8	(−) 7.6	(−) 10.2	Mishra and Agrawal (2015)
	cv HUM- 6		(−) 19	(−) 24.8	(−) 3.8	NS	
Phaseolus vulgaris L...	cv S156	60 ppb (12 h mean)	(−) 38	(−) 52.6	−		Flowers et al. (2007)
	cv Camellino	165 ppb (for 3 h)	(−) 36	(−) 26	−		Guidi et al. (2009)
Oryza sativa L.	cv Saurabh 950	35 ppb (8 h)	(−) 29.3	(−) 12.9	−	(−) 19.4	Rai and Agrwal (2008)
	cv NDR 97		(−) 28.3	(−) 18.8	−	(−) 16.6	
	cv SY 63	13.8–74.2 ppb (7 h d^{-1})	(−) 27.1	(−) 33	−	NS	Pang et al. (2009)
	cv WY J3		(−) 14.8	NS	−	NS	
Triticum aestivum L.		30–119 ppb	(−) 40	(−) 31	−		Feng et al. (2009)
Triticum aestivum L.	cv Sonalika	47.3 ppb (12 h mean)	(−) 15.5	NS	−	(−) 5	Sarkar et al. (2010)
		57.2 ppb (12 h mean)	(−) 28.4	(−) 23.8	−	(−) 7.4	
		67.2 (12 h mean)	(−) 36.2	(−) 38.1	−	(−) 10.7	
	cv HUW 510	47.3 ppb (12 h mean)	(−) 31	NS	−	(−) 11.12	
		57.2 ppb (12 h mean)	(−) 46.21	(−) 22.6	−	(−) 14.6	
		67.2 (12 h mean)	(−) 51.51	(−) 28.7	−	(−) 16.8	

(continued)

Table 3.2 (continued)

Plants/Cultivars		O₃ concentration (ppb)	Percentage change (%)				Reference
			Photosynthetic rate (Ps)	Stomatal conductance (gs)	Internal CO_2 (Ci)	Fv/Fm	
Triticum aestivum L.	cv M234	42 ppb (8 h mean)	(–)27	(–) 20	–	(–) 5.4	Rai et al. (2007)
	cv PBW 343	50.2 ppb (8 h mean)	(–) 19	(–) 33.2	–	(–) 5.9	Rai and Agrawal (2014)
	cv M 533		(–) 18.4	(–) 43.6	–	NS	
Triticum aestivum L. (3 cultivars)		72 ppb (8 h mean)	(–) 20–22	(–) 7–24	–	(–) 9–17	Wahid (2006a)
Triticum aestivum L. (20 cultivars)		82 ppb (7 h mean)	(–) 24	(–) 8	–		Biswas et al. (2008)
Triticum aestivum L. (12 cultivars)		100 ppb (7 h mean)	(–) 36.9	(–) 11.1	–		Biswas et al. (2008)
Hordeum vulgare L. (3 cultivars)		71 ppb (6 h mean)	(–) 13–21	(–) 6–12	–		Wahid (2006b)
Triticum aestivum L. FH-8203		4612 ppb.h (AOT40)	(–) 36	(–) 15	–	–	Adrees et al. (2016)
FH-7096		291.5 ppb (hourly mean)	(–) 38	(–) 60	–	–	
Vigna radiate L. NM-2006		3110 ppb.h	(–) 48	(–) 52.5	–	–	
NM-2011		350.3 ppb (hourly mean)	(–) 37	–	–	–	
Pisum sativum L. Climax		5120 ppb.h	(–) 42	(–) 36	–	–	Adrees et al. (2016)
Peas-09		345.9 ppb (hourly mean)	(–) 42	(–) 36	–	–	

Species	Dose					Reference
Tagetes erecta Linn	6.4 ppm.h	(−) 34[a]	(−) 27[a]	(−) 12[a]	−	Yang et al. (2016)
		(−) 16[b]	(−) 32[b]	−	−	
	8.6 ppm.h	(−) 67[a]	(−) 50[a]	(+) 15[a]	−	
		(−) 42[b]	(−) 67[b]	−	−	
Rosa sinensis Jacp.	AOT40 + 60	(−) 30	(−) 52	(+) 16	−	
	AOT40 + 35	(−) 45	(−) 58	(+) 9	−	
Cotinus coggygria Scop.	Ambient +60 ppb	(−) 22[a]	(−) 29[a]	−	−	
		(−) 17[b]	−	(+) 15[b]	−	
	Ambient +120 ppb	(−) 56[a]	(−) 52[a]	−	−	
		(−) 22[b]	(−) 22[b]	(+) 18[b]	−	
Fagus crenata L.	61.5 ppb	(−) 33.7	(−) 25.8	(+) 54.7	−	Hoshika et al. (2013)

[a]I Sampling; [b]II Sampling

degradation of D1 protein. An imbalance between degradation and de novo synthesis can lead to damage to the photosynthetic apparatus (Yamamoto and Akasada 1995). The light reaction system of photosynthesis is found to be more stable than the dark reaction system under O_3 exposure (Heath 2008). Reichenauer et al. (1997) have shown that Fv/Fm ratio did not vary significantly between *Populus nigra* grown at ambient and elevated (150 ppb) O_3 concentrations, although reduction in photosynthetic yield was significant. These observations suggest that there was no reduction in photochemical capacity (Fv/Fm) upon O_3 exposure, the dark reaction of photosynthesis is impaired by O_3 before any effect on light reaction is detected (Reichenauer et al. 1997). Similar effects of O_3 stress on dark reaction of photosynthesis were observed by D'Hease et al. (2005) and Ranieri et al. (2000). Pellegrini et al. (2011) reported that in *Liriodendron tulipifera*, carboxylation efficiency (V_{cmax}) was reduced by 11% and was the initial cause of impairment of photosynthesis. The detrimental effect of O_3 on carboxylation efficiency was further confirmed by Pellegrini (2014) in *Telia americana*, where a reduction of 35% was recorded upon exposure to O_3 concentration of 120 ppb ($5hd^{-1}$) for 45 days.

Singh et al. (2009) showed that in two cultivars of *Glycine max* exposed to elevated O_3 concentrations of 70 and 100 ppb, inhibition of light reaction may not be completely responsible for reduction in photosynthesis and the suppression of carboxylation efficiency can be correlated to other mechanisms such as suppression of Calvin cycle (Guidi and Degl'Innoenti 2008). In this study, decrease in the quantum yield of PS II may be a mechanism to down regulate photosynthetic electron transport so that ATP and NADPH production is maintained in equilibrium with decreased demand from Calvin cycle in O_3 treated leaves (Liu et al. 2015; Guidi et al. 2001). This kind of result was also obtained by Reichenaeur et al. (1997) who indicated that impairment of dark reaction will lead to an inhibitory effect on light reaction.

Chlorophyll content also plays an important role in determining the photosynthetic yield of the plants. Reduction in chlorophyll content upon O_3 treatment is well cited in literature (Pellegrini 2014; Singh et al. 2009; Kollner and Krause 2003). O_3 treatment may prevent chlorophyll synthesis leading to a net decline in chlorophyll concentration (Castagna et al. 2001). In addition to this, O_3 generated ROS affect the permeability of membrane bound organelles such as chloroplasts leading to destruction of photosynthetic pigments (Rai and Agrawal 2014). Reduction in chlorophyll content as such is indicative of effect of O_3 stress on chlorophyll binding proteins of light harvesting complexes (LHCs). Pellegrini (2014) observed that in O_3 exposed *Telia americana* saplings, Chl a/Chl b ratio showed a reduction of 49% after the end of fumigation, which indicates the reduction in light harvesting complexes of PS II (LHCII). Further it was observed that total chlorophyll/carotenoid ratio decreased by 25% at the end of the fumigation, which suggested that during prolonged O_3 exposure, the plants required to invest in an enhancement of photo protective de-excitation pathways mediated by carotenoids (Pellegrini 2014). A decrease in the value of this ratio along with a consequent increase in carotenoid content is an early stress indicator. This result confirmed the findings of Pellegrini et al. (2011) and established that O_3 treated leaves enhance the need of carotenoid mediated photoprotection and induces a partial breakdown of chlorophylls.

Calatayud and Barreno (2004) reported significant declines of 61.2% and 56.5% in chlorophyll a and chlorophyll b, respectively in *Lactuca sativa* L. var. Valladolid at O_3 concentration of 83 ppb (12 h exposure). At 39 ppb O_3 (12 h exposure), the reductions in chlorophyll a and chlorophyll b were 22% and 17.5%, respectively (Calatayud and Barreno 2004). Calatayud and Barreno (2004) suggested that the significant decrease in chlorophyll a and b and maintenance of low concentration of chlorophyll in leaves was a general feature of plants subjected to oxidative stress due to O_3. It has been suggested that reduction in the plant's pigment content is an adaptational strategy to protect PS II from photooxidation through a reduction in the number of light harvesting antennae (Pellegrini et al. 2011; Rainieri et al. 2001).

Chlorophyll destruction shows alteration with the developmental stage in many cases. Studies have shown that the cell destruction is higher during early developmental stages when the defense mechanisms are insufficient to scavenge the generated ROS, however, during the later stages, O_3 exposed plants get acclimatized to stress conditions and thus chlorophyll loss is lower (Singh et al. 2009). Sakaki et al. (2008). suggested that photosynthetic pigment destruction could be related to superoxide ions that accumulate in O_3 stressed leaves. Guidi et al. (2000) exposed 14 Italian cultivars of *P. vulgaris* to single pulse of O_3 (150 ppb for 3 h) and observed a significant reduction in total chlorophyll content in most of the cultivars and a significant positive correlation was found between chlorophyll content and visible injury symptoms. Sensitive varieties of *P. vulgaris* exposed to 40–50 ppb O_3 for 75–135 min showed higher O_3 induced chlorophyll loss than the tolerant ones (Guzy and Heath 1993). Chaudhary and Agrawal (2014) also recorded significant reductions in chlorophyll content in *Trifolium alexandrium* L. cv Warden grown at elevated O_3.

Reduction in RuBisCO activity also plays an important role in O_3 induced reduction in photosynthetic rate (Bagard et al. 2008; Wittig et al. 2007; Inclan et al. 2005; Gaucher et al. 2003; Dizengremal 2001). Feng et al. (2016) studied the response of five modern wheat cultivars exposed to 1.5 times the ambient concentration of O_3 and reported that significant losses in RuBisCO carboxylation efficiency and RuBP 1,5 bisphosphate regeneration were responsible for lower photosynthetic rates. RuBisCO is regarded as a key protein and is made up of two types of subunits designated as large subunit (LSU) encoded by chloroplastic DNA and a small subunit (SSU), encoded by nuclear DNA. The reduction in RuBisCO activity may be attributed to decline in RuBisCO quantity which may be due to increased rate of protein degradation or inhibition of protein synthesis (Heath 2008). Decreased levels of mRNA transcripts rbc(S) (small subunit) and rbc(L) (large subunit) were observed in the leaves of O_3 treated leaves (Pelloux et al. 2010). Sarkar et al. (2010) observed reductions in the abundance of LSU and SSU in two cultivars of rice, Malviya Dhan 36 and Shivani through one dimentional protein profiling. Transcription of chlorophyll a/b binding protein (cab) and glyceraldehyde 3 phosphate dehydrogenase (gap A and gap B) was also reduced in O_3 exposed plants (Glick et al. 1995).

Goumenaki et al. (2010) reported reduction in net CO_2 fixation in two lettuce varieties (*Lactuca sativa* L. cv Paris and Grenada) exposed to O_3 concentration of 100 ppb (8 hd^{-1}), and this Island was linked to the decrease in the amount of

RuBisCO LSU and SSU subunits. It was further observed that under O_3 stress, the reduction in the transcript abundance of rbcS was more as compared to rbcL (Goumenaki et al. 2010). In addition, O_3 also alleviated the endogenous level of inhibitor of RuBisCO activity, 2-carboxtarabinitol 1-phosphate (CA1P) (Khan et al. 1999). O_3 stress also brings about changes in the ultrastructure of chloroplasts which can be observed as reduction in chloroplast size and disintegration of thylakoid membranes (Gunthardt- Georg et al. 2000). *Zea mays* leaves exposed to 350 ppb O_3 for 5 h for 3 weeks registered an increase in several thylakoid membrane associated proteins, while D1 protein associated thylokoid membrane declined (Pino et al. 1995). Godde and Buchhold (1992) reported variations in the turnover rate of D1 reaction centre polypeptides of PS II in *Picea abies* exposed to 200 ppb O_3.

6.2 Biophysical Variations Upon O_3 Stress

Although several studies have proved that the decline in photosynthetic rate in the plants exposed to O_3 stress is associated with damage to photosynthetic machinery (Feng et al. 2016; Guidi and Degl'Innocenti 2008), biophysical parameters such as stomatal conductance (gs) and internal CO_2 (Ci) also play significant roles in determining the photosynthetic yield and sensitivity of O_3 exposed plants. As discussed above, the damaged photosynthetic machinery reduces the carboxylation efficiency which increases the concentration of Ci in the substomatal cavity, resulting in reduced gs (Fiscus et al. 1997; Mckee et al. 1997; Paoletti and Grulke 2005). Biswas et al. (2008) analyzed the response of 20 varieties of wheat to elevated O_3 and observed that the sensitivity to O_3 in wheat cultivars progressed with the year of release due to higher gs and lower levels of antioxidant capacity of the modern cultivars. *It has been shown that Ps can be regulated by stomatal/non stomatal factors* (Zheng et al. 2014).

Niu et al. (2014) aimed to analyze the role of gs in reducing photosynthetic rates upon O_3 exposure through their experiments on *Cinnamomum camphora* seedlings. The seedlings were exposed to ambient (AA) as well as elevated (AA + 60 and AA+ 120 ppb) O_3 for $8hd^{-1}$ for 2 years (Niu et al. 2014). Observations recorded at the end of the growing season showed reductions of 13% and 25.3% in *Ps* under (AA+ 60) and (AA + 120) ppb O_3 treatments, whereas gs did not show any significant variation with respect to O_3 treatment. These results clearly indicate that reductions in *Ps* of *C. camphora* cannot be attributed to its stomatal behavior. In the same experiment, it was observed that the carboxylation efficiency (V_{cmax}) and electron transport (J_{max}) of RuBisCO decreased significantly in plants grown under elevated O_3 treatment (Niu et al. 2014). Decline in RuBisCO quantity and activity may be responsible for the decline in V_{max} (Dann and Pell 1989), while reduced J_{max} indicates disruptive electron transport under elevated O_3 stress (Warren et al. 2003). Niu et al. (2014) observed a constant ratio of J_{max}/V_{cmax} which suggests that decline in J_{max} may be a result of reduced V_{cmax}. The result of this experiment proved that O_3 induced degradation and deactivation of RuBisCO as well as its feedback inhibitory effect on electron transport system may be the primary cause of reduction in *Ps* rather than the stomatal conductance (Niu et al. 2014). Zhang et al. (2011) also recorded similar response in deciduous broadleaf *Liriodendron chinensis* (Hemsl) Sarg seedlings where reduction in *Ps* was independent of stomatal closure. It was predicted that stomatal closure was not a direct effect of O_3 but a response to increased *Ci* as a result of increased carbon assimilation (Zhang et al. 2010).

Watanabe et al. (2014) studied the effect of O_3 fumigation (60 ppb) on *Ps* in leaves of Monarch birch (*Betula maximowicziana*) and reported a parallel decline in *Ps* along with *gs*, while *Ci* remained constant or even increased suggesting that stomatal closure may only be a downward regulation response rather than a cause of decreased *Ps*. Similar observation was also recorded by Yang et al. (2016) in marigold (*Tagetus erecta* Linn) exposed to O_3 concentrations of 60 and 120 ppb above the ambient levels (AA + 60 and AA + 120 ppb). However, in the same experiment, stomatal limitation seemed to be the principal cause for observed decline in *Ps* in smoke trees (*Cotinus coggygria* Scop.), while in rose (*Rosa chinensis* Jacp.) increased antioxidant capacity was able to sustain the reductions in *Ps* (Yang et al. 2016).

Decreased *gs* is a kind of protective measures adopted by stressed plants as it minimizes the O_3 flux into the plants (Sitch et al. 2007; Wittig et al. 2007). In the expected CO_2 rich and warmer atmosphere of near future (IPCC 2013), plants may show a tendency to reduce stomatal conductance and thus indirectly alleviate the O_3 damage. Hoshika et al. (2013) studied the role of stomatal conductance in defining the avoidance strategy of plants using data on the effects of ambient and elevated O_3 on *Ps* and *gs* in Siebold's beech (*Fagus crenata*) in a coupled photosynthesis-stomatal model. Measurements in the gas exchange parameters were taken in the month of June, August and October and it was found that *Ps* in both O_3 treatments was maximum in June and decreased over time. As compared to the ambient O_3, *Ps* reduced significantly by 34% in August and October, but not in June, whereas *gs* reduced significantly by 21% and 26% in June and August, respectively and remained unaffected during October (Hoshika et al. 2013). Hoshika et al. (2013) also reported a decline in *gs* with no effect on *Ps* of *F. crenata* in June, which clearly indicated the uncoupling of *Ps* and *gs* upon O_3 exposure. The observed reduction in *gs* upon O_3 exposure can be attributed to the modulation of K^+ channels (Torsethaugen et al. 1999) or altered Ca^{2+} homeostasis in guard cells (McAinsh et al. 2002). These observations suggest that *F. crenata* avoids damage to photosynthetic mechanisms by reducing *gs* (Hoshika et al. 2013). However, this adaptational strategy was effective only during June and as the experiment progressed in August and October, this mechanism was not sufficient to prevent reduction in *Ps* under elevated O_3 stress (Hoshika et al. 2013). Uncoupling between *Ps* and *gs* was also reported by Singh et al. (2009) in two soybean varieties (*Glycine max* L. var. PK472 and Bragg) exposed to elevated O_3 concentrations of 70 and 100 ppb ($4hd^{-1}$), wherein the correlation coefficient and regression equation confirmed the uncoupling of the two parameters. Certain modeling studies have also contributed in analyzing the relationship between stomatal O_3 flux and O_3 induced decline in photosynthetic rate in plants (Hoshika et al. 2013; Emberson et al. 2000); Karlsson et al. 2007; Stich et al. 2007; Mills et al. 2010).

Stomatal conductance is an important parameter for differentiating the sensitivities of two crops/cultivars. Singh et al. (2009) observed that gs showed more reduction in O_3 tolerant soybean cultivar Bragg as compared to O_3 sensitive cultivar PK472 under similar O_3 exposure conditions, indicating that Bragg has a better avoidance strategy to O_3 stress. Two way ANOVA test showed that gs varied significantly due to treatments and not to variety, but the interaction between treatment and variety was significant, suggesting that the two varieties responded differently to O_3 treatment (Singh et al. 2009). Koch et al. (1998) reported strong decrease in gs in O_3 resistant poplar hybrid (*Populus maximowizii* x *P. trichocarpa* clone 245) than a sensitive hybrid (clone 388) to O_3 concentration of 300 ppb (6 h daily for 5 days). Similar reductions in gs have also been reported in pumpkin (Castagna et al. 2001), poplar (Guidi et al. 2001), tobacco (Degl'Innocenti et al. 2002) and rice (cvs. Malviya dhan 36 and Shivani) (Sarkar et al. 2015) upon O_3 exposure. et al. (2010) observed reductions in the stomatal conductance of *T. aestivum* L. cvs Sonalika and HUW 510 at O_3 concentration of 36.4–48 ppb. Reductions in gs were also recorded in six cultivars of *Vigna radiate* by Chaudhary et al. (2013) at O_3 concentration ranging between 7.40–57.5 ppb. Wagg et al. (2013), however, found increase in gs in two mesotrophic grassland species, *Dactylus glomerata* and *Ranunculus acris*. The experimental species were exposed to two ranges of 24 h O_3 concentrations ranging between 16.2–33.9 ppb (low) and 72.6–89.5 ppb (high). It was observed that at high O_3 concentration, stomata lost their ability to respond or had reduced response as compared to low O_3 concentration range (Wagg et al. 2013). Thus, this experiment shows an increase in gs upon O_3 exposure, which was higher at high O_3 treatment (Wagg et al. 2013). Table 3.2 shows variations in different physiological paramters upon O_3 exposure.

Stomatal response under O_3 exposure may also be regulated independent of Ci response. Elevated O_3 may directly affect the guard cell functioning, leading to stomatal closure (McAinsh et al. 2002). Paoletti and Grulke (2010) while studying the gas exchange response to static and variable light on snap bean (*Phaseolus vulgaris*), California black oak (*Quercus kelloggii*) and blue oak (*Q. douglassi*) exposed to O_3 (70 ppb 8hd^{-1}) grown in open top chamber reported a delay in stomatal responses which they called as 'Sluggish' responses. Stomatal conductance is genetically regulated so as to maintain the ratio of Ci to ambient CO_2 concentration (Lambers et al. 2008). Sluggish stomatal responses are defined as delay in stomatal movement to changing environmental factors relative to control. Sluggishness increases the time to open (limiting CO_2 uptake) and close stomata (increasing transpirational water loss) (Paoletti and Grulke 2010). Hoshika et al. (2014) examined the effect of O_3 on stomatal dynamics of three common tree species in China (*Ailanthus altissima*, *Fraxinus chinensis* and *Platanus orientalis*) exposed to three levels of O_3 (42, 69 and 100 ppb) and observed that O_3 exposure increased stomatal sluggishness in the following order of sensitivity: *F. chinensis* > *A. altissima* > *P orientalis*. Maier-Maercker (1999) suggested that sluggish behavior of stomata can be explained by delignification of guard cells and subsidiary cells as recorded in O_3 exposed *Picea abies*. However, Paoletti et al. (2009) did not observe any changes in guard cell wall lignifications of O_3 injured *F. ornus* and ascribed slower stomatal response to accelerated senescence in cell physiological processes (Paoletti et al. 2009). The uncoupling of *Ps* and *gs* in plants upon O_3 exposure can be explained by sluggish stomatal responses (Paoletti and Grulke 2005). Hoshika et al. (2016) reported the effect of different light intensities (1500, 1000 and 200 µmol m^{-2} s^{-1}) on O_3 induced stomatal sluggishness in O_3 sensitive S156 genotype of snapbean (*Phaseolus vulgaris*) exposed to 1 h O_3 concentration of 150 ppb and observed that sluggishness was significant only at high light intensity (1500 µmol m^{-2} s^{-1} PPFD) and did not occur at lower light conditions (1000 and 200 µmol m^{-2} s^{-1}).

The stomatal sensitivity towards O_3 stress is smaller in young leaves (Bernacchi et al. 2006) and varies widely between different varieties of same crop (Morgan et al. 2003). Hoshika et al. (2014) suggested that O_3 induced stomatal behavior not only depends upon the amount of O_3 entering the leaves, but also on plant's capacity for biochemical detoxification and repair. Vahisalu et al. (2010) observed that applying 250 ppb O_3 led to 40% reduction in *gs* in wild type (WT) *Arabidopsis* Col-0 plants within 5–10 min exposure which was correlated with the O_3 induced ROS generation. ROS produced under O_3 stress activates OST1 gene of the guard cells which results in the activation of S-type anion channels, resulting in efflux of anions and K$^+$ ions from the guard cells leading to stomatal closure.

7 Effect on the Antioxidant Pool

To avoid the potential damage to the cellular components caused by O_3 induced ROS, the balance between production and elimination of ROS at intracellular level must be competently maintained. This equilibrium is efficiently carried out by enzymatic and non enzymatic antioxidants (Yadav et al. 2004; Caregnto et al. 2013; Mittler et al. 2004). The enzymatic components comprise several antioxidant enzymes, such as superoxide dismutase (SOD), catalase (CAT), glutathione peroxidase (GPX), guaiacol peroxidase (POX) peroxiredoxins (Prxs), and enzymes of the ascorbate-glutathione (AsA-GSH) cycle, such as ascorbate peroxidase (APX), monodehydroascorbate reductase (MDHAR), dehydroascorbate reductase (DHAR), and glutathione reductase (GR) (Sharma et al. 2012; Mittler et al. 2004). Nonenzymatic components include the major cellular redox buffers ascorbate (AsA) and glutathione (GSH) as well as tocopherol, carotenoids and phenolic compounds (Caregnato et al. 2013; Sharma et al. 2012; Mittler et al. 2004; Gratão et al. 2005). Exposure to ambient and elevated O_3 levels results in increased activities of SOD, APX, CAT, POD and GR (Liu et al. 2015; Ueda et al. 2013) as well as enhanced biosynthesis of ascorbate and glutathione (Yendrek and Ainsworth 2015; Dumont et al. 2014). Biswas et al. (2008) observed an increment of 46% in POD activity and a significant reduction of 14% in the foliar ascorbate content in 20 winter wheat cultivars (released in past 60 years in China), exposed to O_3 concentration of 82 ppb (7hd^{-1}) for 21 days. Rai and Agrawal (2014) studied the response of two cultivars of wheat, *T. aestivum* M 533 and PBW 343, exposed to O_3 concentration in the range 49.6–56.4 ppb during vegetative and reproductive stages and reported high POD activity during both the stages in both cultivars, whereas ascorbic acid content increased at both the ages in PBW 343, but in M 533, it increased significantly only during the vegetative stage and showed a reduction during the reproductive age. An increase in POD activity of both the cultivars upon O_3 exposure suggests higher production of H_2O_2 under oxidative stress, whereas increment in ascorbic acid content depicts higher detoxification capacity at earlier stages of development to delay the O_3 induced senescence during the vegetative stage (Rai and Agrawal 2014).

Plants posses a wide range of responses at the biochemical level which assist them to cope with O_3 induced oxidative stress. Differential response of the plant species/varieties to O_3 stress can be attributed to the variations in the constitutive and inducible levels of antioxidants. Many studies have shown that plants with high levels of antioxidants, either constitutive or induced are more resistant to oxidative damage (Wang et al. 2013; Singh et al. 2010). The genotypic variations observed between the different varieties of the same species are dependent on the efficiency of ROS scavenging system to maintain the cellular redox steady state of the leaf tissue (Castagna and Ranieri 2009; Giacomo et al. 2010). Ascorbate pool within the apoplastic/symplastic regions of the leaves plays an important role in defining the sensitivity/resistivity of a plant cultivar (Frei et al. 2012). The protective role of AsA as ROS scavenger was first demonstrated by enhanced O_3 sensitivity shown by *Arabidopsis thaliana* mutants deficient in AsA synthesis (Conklin et al. 1996). Cargenato et al. (2013) working on two Brazilian cultivars of *P. vulgaris* L. Irai and Fepagro 26 exposed to O_3 dose of 122.6 ppb.h observed that Irai has a higher constitutive AsA level than Fepagro 26 and O_3 exposure does not modify the level of foliar AsA pool in Irai as compared to Fepagro 26. This result indicates that Irai is capable of sustaining a higher AsA mediated antioxidant activity by maintaining a balance between the extra and intracellular supply of reduced AsA, at a level sufficient enough to detoxify the ROS generated under O_3 stress (Caregnato et al. 2013). In this study, Caregnato et al. (2013) tried to correlate the AsA regeneration rate with O_3 flux and observed that in Fepagro 26, in spite of the fact that O_3 exposure increases the concentration of AsA, its antioxidant capacity to respond to O_3 flux is less effective than that of Irai (Caregnato et al. 2013). The results of this experiment suggest that AsA antioxidant capacity per unit O_3 flux is less in Fepagro 26 than in Irai, due to which more O_3 molecules are free to be generated in the foliar apoplastic region rendering Fepagro 26 more sensitive to O_3 than Irai (Caregnato et al. 2013).

Singh et al. (2010) also observed higher levels of ascorbate in resistant cultivar of soybean (*G.max* L. Bragg) as compared to sensitive cultivar PK472 when exposed to 70 and 100 ppb ($4hd^{-1}$) O_3 concentrations. Maddison et al. (2002) have shown that the tolerance to O_3 increased in an O_3 – sensitive genotype of *Raphanus sativus* L, when its L-acorbate content was increased by feeding hydroponically cultivated plants with the biosynthetic precursor, L-galactano-1,4-lactone. Several other investigations have also shown that O_3 tolerant genotypes have higher ascorbic acid content than the sensitive ones. Rai and Agrawal (2014) indicated variations in the ascorbic acid content of two wheat cultivars (PBW 343 and M 533) in response to O_3 concentration of 53.2 ppb. Feng et al. (2010) also found that ascorbate increased by 33.5% in Y16, a tolerant cultivar, than Y2 (26.7%), a sensitive cultivar, of wheat at elevated O_3 concentration of 66 ppb. Sarkar et al. (2015) reported that increments in ascorbic acid content was higher (24.5%, 35.3% and 62.1%) in O_3 tolerant rice cultivar Malviya Dhan 36 as compared to O_3 sensitive cultivar Shivani, in which ascorbic acid increments were 8.6%, 19.6% and 31.3% at O_3 concentrations of 52.2, 62.2 and 72.2 ppb, respectively. Higher levels of ascorbic acid in leaf extracts of O_3 tolerant genotypes of snap beans (Burkey et al. 2003) and *Plantago major* (Zheng et al. 2000) have also been reported.

The reduced ascorbate pool is replenished through the AsA- GSH cycle which has an important role in eliminating highly toxic H_2O_2 generated via O_3 stress (Rao et al. 1995). The AsA component of this cycle is responsible for scavenging the O_3 induced ROS while GSH component works to regenerate/replenish the reduced ascorbate pool (Foyer and Noctor 2005). The ratio of AsA/DHA (reduced ascorbate/oxidized ascorbate) and GSH/GSSH (reduced glutathione/oxidized glutathione) may function as signals for regulation of antioxidant mechanism (Mittler 2002). Wang et al. (2013) studied the response of two rice cultivars (Oryza sativa) SY63 (O_3 sensitive) and WXJ14 (O_3 resistant) exposed to elevated O_3 (1.5 times the ambient concentration) and observed that AsA-GSH cycle was more efficient in O_3 resistant cultivars. Average concentration of AsA was 35.6% higher in WXJ14 as compared to SY63, whereas DHA content was up by 70.2% in SY63 as compared to WXJ14. It was observed that AsA/DHA increased during early growth stages and decreased during later growth stages in SY63, while WXJ14 showed a reverse trend (Wang et al. 2013). GSH/GSSG ratio in SY63 decreased throughout the growth period and reduced significantly during later growth stages whereas in WXJ14, reduction was more prominent during early growth stages (Wang et al. 2013).

Based on the levels of AsA regenerating enzymes MDHAR and DHAR, Wang et al. (2013) concluded that the early increase in AsA/DHA ratio under elevated O_3 in SY63 was due to AsA recycling capability rather than AsA biosynthesis, while in WXJ14 the decrease in later growth stage was attributed to reduction in AsA biosynthesis rather than its recycling capability. In SY63, the DHA and GSSG contents accumulated throughout leading to reductions in AsA/DHA and GSH/GSSG ratio during later growth stages upon exposure to elevated O_3 indicating that AsA and GSH were consumed more rapidly than in WXJ14 (Wang et al. 2013). In WXJ14, the upregulation of GSH/GSSG during early growth stage helped the cultivar in providing protection against O_3 stress during later growth stages. This experiment proves the role of GSH pool in defining the genotypic sensitivity/resistivity of plants to O_3 stress. Dumont et al. (2014) further strengthened the role of AsA and GSH pool in determining the genotypic sensitivity/tolerance. O_3 significantly increased the total ascorbate and glutathione contents in three Euramerican poplar genotypes, (*Poplar deltoids* Bartr x *Populus nigra* L.) 'Carpaccio', 'Cima' and 'Robusta' differing in O_3 sensitivities, with the most resistant genotype 'Carpaccio' showing an increase up to 70% in total ascorbate and glutathione contents followed by 'Cima', whereas the most sensitive genotype 'Robusta' seemed unable to regenerate AsA from oxidized ascorbate (DHA) (Dumont et al. 2014).

The enzymatic antioxidants also play significant roles in scavenging the ROS generated upon exposure to O_3 stress. SOD acts as a major scavenger of O_3 generated $O_2^{\circ-}$ in the apoplast and result in formation of H_2O_2, which in turn is acted upon by different PODs present in the plants (Barcelo et al. (2003). Thus, the activities of SOD and PODs are often coupled together. Wang et al. (2014) reported a significant increase in the apoplastic SOD in O_3 resistant wheat cultivar Y16, exposed to elevated O_3 (1.5 times ambient O_3). However, during the later growth period, apoplastic SOD activity declined by 21.3% in O_3 tolerant Y16 and by 39.8% in O_3 sensitive Y19 cultivar under elevated O_3 as compared to ambient O_3. Ueda et al. (2013), however, observed a variable response in the cytosolic and organelle SODs in a Japanese rice cultivar (*Oryza sativa* L. var. Koshihikari) exposed to O_3 concentration of 150 ppb for 6 h. In this experiment, the peroxisomal, mitochondrial and chloroplastic SODs down regulated their gene expression levels significantly while the cytosolic SOD maintained their expression level, which suggest that organelle SODs play more important role than the cytosolic ones under elevated O_3 stress conditions (Ueda et al. 2013). Sarkar et al. (2015) reported an increased induction of SOD in both the rice cultivars Malviya Dhan 36 and Shivani, along with increased activities of CAT and POD, which indicates towards an efficient detoxification of H_2O_2 produced under O_3 induced oxidative stress (Sarkar et al. 2015). Reductions of 27% and 38.3% in POD activity were recorded in two wheat cultivars, Y16 and Y19 during the later growth stage upon exposure to elevated O_3 as compared to ambient O_3 (Wang et al. 2014). A comparison between POD activity of recent wheat cultivar PBW 343 and an old cultivars M 533 (O_3 concentration: 53.2 ppb) (Rai and Agrawal 2014) and M234 (O_3 concentration: 41 ppb) (Rai et al. 2007) showed that recent cultivar (PBW 343) showed a higher magnitude of induction of POD activity as compared to older cultivars M234 and M533, suggesting that recent cultivars had a better H_2O_2 scavenging capacity that the older ones. Cho et al. (2008) reported increased amount of CAT and POD transcripts in rice seedlings under elevated O_3 exposure.

APX is one of the most important ROS scavenging antioxidant enzyme whose efficiency is attributed to its presence in different cellular compartments and its high affinity for H_2O_2 (Castagna and Rannieri 2009; Sharma et al. 2012). Pasqualini et al. (2001) have associated the high tolerance of O_3 resistant tobacco cv. Bel B with a greater antioxidant enzyme activity. Bandurska et al. (2009) have correlated APX activity with O_3 injury symptoms in two cultivars of tobacco and found that O_3 resistant cultivar (Bel B), which did not show any O_3 injury symptom has high APX activity as compared to O_3 sensitive cultivar Bel W3, which showed significant O_3 injury symptoms. These results suggest that activation of antioxidant defense system in O_3 tolerant cultivar prevented the appearance of O_3 induced injury symptoms (Bandurska et al. 2009). Along with APX, gluicol peroxidase (GuPx) also showed significant variations under O_3 stress. A highly positive and significant correlation between GuPX activity and O_3 concentration in both sensitive and resistant cultivars of snap beans (*Phaseolus vulgaris*) (Burkey et al. 2000) and tobacco (Bandurska et al. 2009) was reported. However, unlike APX, GuPX activity cannot be utilized for differentiating between O_3 sensitive and resistant cultivars.

Chernikova et al. (2000) studied O_3 tolerance in two cultivars of soybean (*Glycine max*) cv Essex (O_3 tolerant) and Forrest (O_3 sensitive) on the basis of response of their antioxidant content, under exposure to mean O_3 concentration of 62.9 ppb. The result of this experiment clearly depicted that O_3 tolerant Essex cultivar had less oxidative damage and showed higher levels of GR (30%), APX (13%) and SOD (45%) activities as compared to O_3 sensitive Forrest (Chernikova et al. 2000). However, increment in the activity of GuPX was higher in Forrest (62.8%) compared to Essex (39 23%) under similar O_3 exposure conditions (Chernikova et al. 2000). Scebba et al. (2006) found greater induction of POD activity in O_3 resistant *Trifolium repens* as compared to O_3 sensitive *T. pretense* upon exposure to acute dose of 150 ppb O_3 for 3 hrs. Chaudhary and Agrawal (2013) studied intra -specific variations in six cultivars of *T. alexandrium* exposed to mean O_3 concentration of 57.5 ppb, with respect to activities of SOD, POD and APX and found that O_3 tolerant cultivars showed higher induction of enzyme activities confirming their ability to provide higher level of protection against oxidative stress.

Chernikova et al. (2000) further observed that cytosolic APX (cAPX) was more responsive towards O_3 treatment as compared to stromal APX (sAPX), the activity being more prominent in O_3 tolerant cultivars. This fact was strengthened by Torsethaugen et al. (1997), who showed that transgenic tobacco (*Nicotiana tabaccum*) with enhanced sAPX activity (in chloroplasts) does not necessarily lead to O_3 tolerance. In case of O_3 stress, where H_2O_2 first appears in the cytosol, the role of cAPX becomes more important. However, when plants experience stress due to internal factors, which lead to the disruption of chloroplast electron transport, sAPX assumes a significant role in stress tolerance. There are reports of increased O_3 sensitivity in *N. tabaccum* with suppressed cAPX activity, but no improvement in O_3 tolerance was observed upon over expression of cAPX (Orvar and Ellis 1997). This observation can be explained by the fact that O_3 tolerance does not depend solely upon APX activity but also upon other antioxidant enzymes like SOD and GR and on adequate regeneration of ascorbic acid (Chernikova et al. 2000).

Sarkar et al. (2015) reported significant increments in major antioxidant enzymes in two rice cultivars Malviya Dhan 36 and Shivani upon exposure to elevated O_3. APX showed increments of 19.4, 43.1 and 55.3% in cv Malviya Dhan 36 and 13.7, 59.2 and 72.4% in Shivani at mean O_3 concentrations of 52.2, 62.2 and 72.2 ppb, respectively. Similar increments in both the cultivars were recorded for GR (15, 60 and 90% in Malviya Dhan 36 and 23.3, 54.2 and 104% in Shivani) under similar O_3 exposure conditions (Sarkar et al. 2015). Since APX and GR are important constituents of ascorbate-glutathione cycle, their increased activities may promote the regeneration of ascorbic acid and can be regarded as a strategic management of O_3 induced oxidative stress. Similarly, induction of APX under O_3 induced stress was also reported in different cultivars of wheat (Sarkar et al. 2010; Agrawal et al. 2002), rice (Rai and Agrawal 2014; Wang et al. 2013) and maize (Singh et al. 2014). Mishra and Agrawal (2015) also reported enhanced antioxidative enzyme activity in two cultivars of *Vigna radiate* (HUM-2 and HUM-6) upon exposure to mean 8 h O_3 concentration of 68.9 ppb with HUM-6 showing more efficient enzymatic defense mechanism as compared to HUM-2.

Wang et al. (2014) emphasized that differential antioxidant deployment at different developmental stages were responsible for the development of resistant strategies towards O_3 stress. Evaluating the antioxidative response of two wheat cultivars Yangmai 16 (Y16) and Yonnong 19 (Y19), Wang et al. (2014) observed that elevated O_3 (1.5 times ambient) induced a decrease in activities of antioxidant enzymes (SOD, POD and APX) during later growth stage, indicating higher sensitivity of Y19 during later grain filling stage. However, in case of Y16, up regulation of SOD and APX at the booting (early growing) stage facilitated the delay in the effect of O_3 on pant senescence, providing more tolerance to Y16 towards O_3 stress as compared to Y19 (Wang et al. 2014). Age specific response of antioxidants was also reported in five wheat cultivars (Yangmai 16, Yangmai 15, Yangfumai 2, Yannong 19 and Jiaxing 002) exposed to elevated O_3 (1.5 times the ambient concentration), where significant O_3 effects were recorded only during the mid grain filling stage (Feng et al. 2016).

Biswas et al. (2008) studied the response of 20 wheat varieties released over past 60 year in China and observed that although the recent varieties had a higher capacity to detoxify the O_3 induced ROS, they were more sensitive to O_3 as compared to the older varieties. The higher tolerance of the older varieties was attributed to lower O_3 uptake (gs) as well as higher potential capacity to repair O_3 induced oxidative damage (Biswas et al. 2008). Feng et al. (2016), however, observed that five modern wheat cultivars (Yangmai 16, Yangmai 15, Yangfumai 2, Yannong 19 and Jiaxing 002) upon exposure to elevated O_3 (1.5 times the elevated O_3; M7 for ambient O_3 averaged 52 ppb with a maximum of 110 ppb) showed significant interactions between O_3 and cultivars for antioxidant enzymes but not for gs. Therefore, it was concluded that in the modern wheat varieties, the antioxidant enzymes rather than gs were found to be responsible for differential response to elevated O_3 (Feng et al. 2016).

8   Effects on Metabolites

O_3 stress brings about significant variations in the metabolite contents of the plants (Wang et al. 2013; Singh et al. 2010; Iriti and Faoro 2009). O_3 may cause shifts in the partitioning of assimilates from storage compounds (i.e. starch) to compounds involved in O_3 injury repair response. The carbon metabolism becomes more inclined towards synthesis of compounds such as lipids, organic acids, phenol and structural carbohydrates, which support the plants in escalating their tolerance/resistivity towards O_3 stress (Friend and Tomlinson 1992). Carbohydrates, proteins and phenols are important metabolites, affected by O_3 stress. Studies have confirmed that plants growing under O_3 stress convert mobile carbohydrates to less mobile secondary metabolites (Hain 1987).

O_3 treated plants show a decline in their foliar carbohydrate content (Rai et al. 2011; Iglesias et al. 2006; Kollner and Krause 2003). Reduction in the carbohydrate content may be a consequence of diminished photosynthetic rate or allocation of carbon metabolism pathway through modification of source- sink balance (Andersen 2003) or carbon allocation processes (Drogoudi and Ashmore (2002). Grantz and Young (2000) showed that O_3 directly affects the phloem transport with consequent inhibition of translocation to roots. Singh et al. (2010), however observed a contrasting pattern of biomass allocation i.e. more biomass was allocated towards the roots of two soybean cultivars (*G. max* cv PK472 and Bragg) exposed to elevated O_3 concentrations of 70 and 100 ppb. This unexpected behavior of biomass allocation in soybean is attributed to the specific feature of nitrogen in the plants. The phenomenon of nitrogen fixation utilizes extra energy which is provided by allocating more biomass towards the roots. Some earlier studies have also reported increased allocation of photosynthates to roots following O_3 exposure (Reiling and Davison 1992; Davison and Barnes 1998).

O_3 stress leads to an increase in the catabolic pathways which allow detoxification processes to remain active through a continuous supply of reducing power (Dizengremel et al. 2008; Gillespie et al. 2011). Further, under O_3 stress, plants divert a large part of their carbon skeleton towards pathways leading to the synthesis of secondary metabolites like phenolic compounds (Castagna and Ranieri 2009). Phenolics provide the plants with specific adaptations to environmental stress and therefore, are essential for plant defense mechanisms (Nakabayashi and Saito 2015). Development of any defense trait is an expensive feature in plants as it diverts the energy (biomass) from growth to defense metabolite production. This feature of plants leads to a condition called 'trade- off' i.e. promoting some metabolic functiong by neglecting the others (Caretto et al. 2013). Under O_3 stress, plants tend to follow the anaplerotic pathway, which tends to reduce Ribulose 1,5 bisphosphste carboxylse/oxygenasse (RuBisCO) activity and enhance Phosphoenolpyruvate carboxylase (PEPc) activity (Dizengremel 2001; Dalstein et al. 2002; Renaut et al. 2009). Gene related studies have shown that photosynthesis related genes were suppressed while those related to catabolic processes were enhanced (Ahsan et al. 2010; Agrawal et al. 2002). In plants, RuBisCO in chloroplasts is the main carboxylating enzyme, while PEPc refixes respiratory CO_2 in the cytoplasm. The inverse relationship between these two enzymes has been well studied in different plant species (Table 3.3). Under stress conditions, the increased activity of PEPc tends to stimulate the shikimate or phenylpropanoid pathway leading to production of flavonoids or phenolic compounds (Cabane et al. 2004; Betz et al. 2009). This diversification of pathway is evident through the observation of O_3 induced increase in gene transcription and activities of large number of enzymes of chloroplast involved in shikimate and phenypropanoid pathways. O_3 induced increase in the phenolic content with a simultaneous decrease in protein content was reported by several workers, which also justifies the shift in the metabolic pathways. Increase in phenol content with a parallel decrease in protein content of plants in response to O_3 stress suggest the diversion of amino acid moieties towards phenol metabolism. According to protein competition model of phenolic allocation, an inverse relationship exists between protein and phenol contents due to a common limiting precursor, phenylalanine (Jones and Hartley 1999). Mishra and Agrawal (2015) reported an inverse correlation between protein and phenol contents in two cultivars of *Vigna*

Table 3.3 Rubisco and PEPc activities (nkat mg^{-1} protein) and ratios of the two enzymes in leaves of trees exposed to elevated ozone

	Plant/Species	Ozone concentration (ppb)	Percent change (%)		Rubisco/PEPc		Reference
			Rubisco	PEPc	(C)	(O)	
1.	*Picea abies* cv Gerardmer	200 ppb (3 months)	53.1 (↓)	980 (↑)	25.6	1.11	Antoni (1994)
2.	*P. abies* cv Istebna	200 ppb (3 months)	53 (↓)	466 (↑)	29.3	2.43	Antoni (1994)
3.	*Pinus halepensis*	200 ppb (3 months)	44.8 (↓)	256.3 (↑)	24.7	3.82	Fontaine et al. (1999)
4.	*Populus tremula x alba*	60 ppb (1 month)	46.5 (↓)	275 (↑)	47.2	6.74	Deschaseaux (1997)
5.	*Fagus sylvatica*	110 ppb (4 months)	49.5 (↓)	36 (↑)	6.9	2.56	Lutz et al. (2000)
6.	*Acer saccharum*	200 ppb (2 months)	31.2 (↓)	130.6 (↑)	3.4	1.01	Gaucher et al. (2003)

(↓) = Decrease; (↑) = increase; (C) = Control; (O) = Ozone- fumigated

radiate grown at 8 h mean O_3 concentration of 68.9 ppb. Such correlations were also recorded in radish and brinjal grown at 8 h mean O_3 concentration of 68.9 ppb (Tiwari and Agrawal 2011), and in wheat (M 234 and PBW 510) grown at 12 h mean O_3 concentration of 50.2 ppb (Rai and Agrawal 2014). Higher phenol content is an adaptive feature of plants under O_3 stress as phenols have the ability to act as an antioxidant due to its free radical trapping properties (Heath 2008). Phenols are recognized as strong targets for O_3 phytotoxicity and get oxidized by peroxidases in presence of H_2O_2 and can be reduced to their parent compounds by non enzymatic reaction with ascorbate (Takahama and Oniki 1997; Sgarbi et al. 2003).

Wang et al. (2014) studied the effect of elevated O_3 (1.5 × ambient) on flavonoid contents of two wheat cultivars Y16 and Y19 and observed that total flavonoid content in Y19 increased significantly by 14% during early growth stage (booting), while during later growth stage, it decreased significantly by 12.5% and 11.8%, respectively in Y16 and Y19. As flavonoids play an important role in scavenging O_3 induced ROS, increased flavonoid content in Y19 during booting stage indicated a self protection response against O_3 stress. Phenolic contents increased significantly (12.96%) in Y16 during early growth stage, but decreased significantly during later growth stage (Wang et al. 2014). Y19, on the other hand showed a continuous decreasing trend throughout its growth period. Results showed that during the earlier growing stages, elevated O_3 increased total flavonoid content in Y19 and total phenolic content in Y16, respectively, while on the later growing stages, elevated O_3 decreased total flavonoid content in the leaves of both the wheat cultivars. The response of these two secondary metabolites suggested that short term O_3 exposure induced production of phenols and flavonoids as ROS scavenging mechanisms, but during the later growth stages, the O_3 induced ROS generation exceeded their scavenging capacities, thus leading to decline in their contents (Wang et al. 2014). In the same experiment, Wang et al. (2014) also studied the response of secondary metabolism related enzyme activities which included Phenylalanine ammonialyase (PAL), Polyphenol oxidase (PPO) and Lipoxygenase (LOX). The results showed that in Y16, PAL activity increased (42.3%) upon elevated O_3 stress during early growth stage, while in Y19, it declined (14.6%) during heading stage. PPO and LOX showed a continuous increasing trend throughout the growth period in both the cultivars (Wang et al. 2014). PAL and PPO activities were regulated by flavonoid and phenol metabolism, respectively, whereas LOX acts as a kind of oxidoreductase (Wang et al. 2013). Results of this experiment indicated that differential response of these three enzymes between the two wheat cultivars can be attributed to the secondary metabolic mechanisms adopted by the two wheat cultivars (Wang et al. 2014).

O_3 generated stress brings about modifications in amino acid fragments and peptide chains, alters the electric charge and increases the susceptibility of proteins to proteolysis (Møller and Kristensen 2004). This results in reduction of overall protein content of plants under O_3 stress (Tiwari and Agrawal 2011; Rai and Agrawal 2014). Reduction in RuBisCO content as discussed above is also an important factor responsible for the observed decline in the protein content of O_3 exposed plants. Radish and brinjal grown at 8 h mean O_3 concentration of 40.8 ppb showed reductions of 15.5% and 16%, respectively (Tiwari and Agrawal. 2011) and carrot showed a reduction of 23.9% at 8 h mean O_3 concentration of 38.4% (Tiwari et al. 2006) as compared to filtered air. Soybean cultivar PK472 showed reductions of 26.4% and 30% and Bragg 22% and 20% in foliar protein contents at 70 and 100 ppb (4 hd^{-1}) O_3 concentrations, respectively (Singh et al. 2010). Reduction in protein contents was higher during later part of the life cycle, suggesting the predominance of senescence related processes upon O_3 exposure (Singh et al. 2010). Agrawal and Agrawal (1990) observed reductions of 39% and 6.8% in protein contents, respectively in *Vicia faba* and *Cicer arietinum*, 30 days after exposure of 99 ppb O_3 (2 hd^{-1}). Variations in the magnitude of reductions in protein content in different cultivars of wheat, Sonalika and HUW 510 (Sarkar et al. 2010) and M533 and PBW343 (Rai and Agrawal 2014) have been reported.

Protein profiling tests have also confirmed the reduction in protein content of plants under O_3 stress. SDS PAGE protein profile done on two rice cultivars Malviya Dhan 36 and Shivani showed that protein content significantly reduced in O_3 injured leaves (Sarkar and Agrawal 2010). Monomeric proteins in the range 54, 35 and 33 kDa reduced significantly, while 15.7 and 13.9 kDa completely disappeared. In NATIVE PAGE, maximum reduction was observed at 335 KDa while other points showed lesser reduction in O_3 injured leaves (Sarkar and Agrawal 2010). Through two dimensional separation of total proteins, it was shown that the total amount of RuBisCO declined in two cultivars of rice, Malviya Dhan 36 and Shivani exposed to O_3 concentration ranging from 52–72.6 ppb (Sarkar et al. 2015). Ahsan et al. (2010), through proteome analysis of soybean under acute O_3 exposure of 120 ppb for 3 days reported that proteins associated with photosynthesis and carbon assimilation decreased while those involved in antioxidant defense and carbon metabolism increased. Agrawal et al. (2002) studied the total protein content of rice seedlings through two dimensional electrophoresis and reported that O_3 caused significant reduction in foliar photosynthetic proteins especially RuBisCO and induced various defense related proteins including a pathogenesis related (PR) class 5 protein, three PR class 10 proteins, APX, SOD, calcium binding protein, etc. O_3 affects the synthesis and degradation of both large and small subunits of RuBisCO (Agrawal et al. 2002). Singh et al. (2014) reported significant reductions in PEPc and RuBisCO proteins in two cultivars of *Zea mays*, DHM117 and HQPM1, with a higher loss observed in DHM117 at elevated O_3 concentration between 80–90 ppb.

9 Effect of Ozone on Nitrogen Metabolism

Nitrogen is an essential element required for the proper growth and development of the plants and acts as an important component of proteins and chlorophyll (Robinson 2005). The most important source of nitrogen in plants is the nitrate (NO_3^-) and ammonium (NH_4^+) ions present in the soil. The process of nitrogen assimilation involves the uptake of nitrogen through roots in form of NO_3^- or NH_4^+ ions. The further steps require an incorporation of the absorbed NO_3^- and NH_4^+ as amino groups into carbon compounds to form amino acids (Larcher 2003; Forde and Lea 2007). These reactions involve enzymes such as nitrate reductase, nitrite reductase and glutathione synthetase. O_3 is known to affect the activities of these enzymes, thus influencing the biosynthesis of amino acids. Many studies have reported that exposure to O_3 increased the concentration of free amino acids and the activity of glutathione synthetase in the needles of *Pinus taeda* L. (Manderscheid et al. 1992), *P. sylvestris* (Kainulainen et al. 2000), *Picea abies* (Holopainen et al. 1997), *Fagus crenata* (Yamaguchi et al. 2007b). However, a reduction in photosynthetic nitrogen use efficiency (PNUE) in *Quercus serrata* seedlings was observed (Watanabe et al. 2007). Yamaguchi et al. (2007a, b) exposed Japanese deciduous broad leaved forest tree species to O_3 at 1, 1.5 and 2 times the ambient O_3 concentrations and treated them with different nitrogen loads. Yaamguchi et al. (2007b) observed that O_3 induced significant reductions in concentration of RuBisCO and total soluble proteins in the leaves. A negative correlation of RuBisCO and total soluble proteins with the concentration of foliar acidic amino acids was observed, suggesting that O_3 enhanced the degradation of proteins such as RuBisCO (Yamaguchi et al. 2007b). Yamaguchi et al. (2007a) studied the PNUE in *Fagus crenata* and observed that the degree of O_3 induced reduction in nitrogen availability to photosynthesis was greater in relatively high nitrogen treatment than that in low nitrogen treatment. This suggests that O_3 not only induces alterations in foliar nitrogen metabolism, but also brings about reduction in availability of nitrogen for photosynthesis in the leaves of the trees (Yamaguchi et al. 2007a). Yamaguchi et al. (2007b) proposed the hypothesis that enzymatic nitrogen assimilation activity and protein degradation rates may be altered by elevated O_3 depending upon the availability of soil nitrogen.

Yamaguchi et al. (2010) emphasized on the role of O_3 on nitrogen metabolism in the leaves of *Fagus crenata* seedlings supplied with nitrogen at 0 (N0), 20 (N20) and 50 (N50) kg ha^{-1} year^{-1} and exposed to 24 h O_3 concentration of 42.7, 63.3 and 83.7 ppb in open top chambers. The experimental results showed that exposure to O_3 significantly increased the relative contents of acidic amino acids in all the nitrogen treated plants while it significantly reduced the concentration of total soluble proteins (TSP) and the ratio of TSP to leaf nitrogen concentration in N50 treatment (Yamaguchi et al. 2010). Increase in the contents of acidic amino acids upon O_3 exposure was also reported in the needle of *P. taeda* (Manderscheid et al. 1992). It was further observed that O_3 induced reduction in soluble proteins was greater under relatively high nitrogen load than under relatively low nitrogen load (Yamaguchi et al. 2010). Based upon this study, Yamaguchi et al. (2010) predicted that O_3 not only reduced allocation of nitrogen to soluble proteins, it also stimulates degradation of proteins in foliage of seedlings grown under high nitrogen load. Yamaguchi et al. (2010) also reported a reduction in the activity of nitrate reductase (NR) in O_3 exposed seedlings of *F. crenata* in N50 treatment, but not in N0 and N20 treatments. O_3 induced reductions in NR activity have also been reported by Agrawal and Agrawal (1999) in snapbeans and Smith et al. (1990) in soybean upon O_3 exposure. It is well established that activity of NR depends upon the rate of photosynthesis (Kaiser and Huber 2001). Yamaguchi et al. (2007a) reported a significant reduction in the rate of photosynthesis by exposure to O_3 at higher nitrogen treatment. These two experimental data suggest that inhibition of NR by O_3 was mainly due to O_3 induced reduction in net photosynthesis rate. Yamaguchi et al. (2010) further reported that there was no significant effect of O_3 on the activity of glutamine synthetase. These results indicate that the biosynthesis of amino acids was not affected by O_3 exposure in the leaves of *F. crenata* seedlings.

It is suggested that exposure to O_3 impairs the resorption of nitrogen from the leaves of forest trees species (Udding et al. 2005). O_3 induced degradation of proteins may reduce the allocation of nitrogen to proteins in the leaves. Yamaguchi et al. (2010), however, reported a significant allocation of nitrogen to soluble protein only at higher nitrogen treatment (N50) and not at lower treatments (N0 and N20). This result can be explained by the fact that the amount of nitrogen supplied to the soil and the soil nitrogen availability to plants of N0 and N20 treatments was relatively low as compared to N50 treatment. The O_3 induced reduction in TSP in N50 treatment (as reported by Yamaguchi et al. 2010) can be attributed to O_3 induced degradation of proteins in leaves of *F. crenata* seedlings. Results obtained through different studies (Yamaguchi et al. 2007a, b, 2010) suggest that elevated O_3 can suppress the role of nitrogen as a fertilizer. As the nitrogen supplement is increased, negative effects of O_3 on nitrogen metabolism on pants can be partially mitigated. However, further studies are required to clarify the effects of O_3 on nitrogen metabolism in plants.

A special group of plants belonging to the family Fabaceae are characterized by the presence of biological nitrogen fixation. This process depends upon the symbiotic association of these plants with nitrogen fixing bacteria generally belonging to the genera *Rhizobia* which are found in specialized structures on roots called nodules. Nitrogen fixation in nodulating legumes is a feature that has evolved during early history of the legume family (Sprent 2007). Several studies have highlighted the sensitivity of economically important legumes such as soybean, clover, pulses, beans, etc. to ground level O_3 (Mills et al. 2007). Hewitt et al. (2016) analyzed a total of 26 studies related to the effect of O_3 on nitrogen fixation and observed that the most common effect of O_3 on the number, size and mass of root nodules, with 17 studies reporting a negative impact of O_3 on root nodulation. Cong et al. (2009) analysed a seasonal exposure of peanuts to O_3 treatment varying between 49 and 70 ppb is sufficient to reduce nitrogen fixation rates as compared to charcoal filtered controls. Hewitt et al. (2014) studies the response of two clover species, *Trifolium repens* cv Crusader and *T. Pratense* cv. Merviot to 24 h mean O_3 concentration varying from 33–60 ppb and observed that both the cultivars showed significant reductions in number of nodules per pot (32% and 36% in Crusader and Merviot, respectively) and mass per nodule (36% and 60% in Crusader and Merviot, respectively). Significant reductions in nitrogen fixation were also reported in both the cultivars of clover upon exposure to elevated O_3 (Hewitt et al. 2014). Cowpea (*Vigna unguiculata* (L) Walp) exposed to 8 h O_3 concentration of 40 and 70 ppb showed significant reductions of 42.30% and 55.57%, respectively in nodule dry weight and 50.68% and 51.7%, respectively in number of nodules (>2 mm) (Umponstira et al. 2009). It was further reported that cowpea showed significant declines of 30.7% and 33.85% in the nitrogenase activity at 40 and 70 ppb O_3 concentration, respectively

(Umponstira et al. 2009). Significant reductions in the number and dry mass of soybean root nodules were also recorded in plants exposed to 100 ppb O_3 concentration (Zhao et al. 2014).

Reduced nitrogen fixation rates in nodulating legumes may be an outcome of decreased root nodule size or number or due to reduced nitrogenase activity (Rees et al. 2005). As assessed through isotopic studies, reduction in nitrogen fixation rates is considered to be a result of reduced availability and translocation of carbon assimilates to the root system (Udvardi and Poole 2013). O_3 not only directly causes reduction in the photosynthetic rate, it also results in diversion of carbon and other resources to above ground growth, defense and repair including synthesis of antioxidants and structural carbohydrates (Wilkinson et al. 2011). Hewitt et al. (2014) suggested that under chronic O_3 exposure, reduced nitrogen fixation may be associated with reduction in root biomass. In case of *T. pratense*, intense negative impacts of O_3 on nitrogen fixation can be avoided due to high stomatal conductance rate and the possible ability to maintain translocation of assimilates to root system (Hewitt et al. 2014).

10 Conclusion

The phytotoxic nature of O_3 was first confirmed by Richards et al. (1958) when O_3 injury symptoms were observed on grape foliage. Since then, several meticulously planned experiments performed by several workers around the globe have proved the negative effects of O_3 on plant performance. O_3 enters the plants through stomata and dissolves in the apoplastic fluid to form ROS, which are the main culprits responsible for causing O_3 induced damage. Plants have their own intrinsic defense machinery, which includes the enzymatic and non enzymatic antioxidant pool of the cell. The degree of O_3 damage incurred by the plants can be correlated to the balance between ROS generated due to O_3 stress and the antioxidant pool active in the cell. When the level of ROS exceeds the antioxidative capacity of the cell, the plants show O_3 induced injury symptoms. Membranes are the first targets of the oxidative action of O_3, wherein the O_3 induced peroxidation of the lipid component of the membranes results in the disorganization of membrane permeability. O_3 severely affects the photosynthetic processes not only by disrupting the electron transport process but also by reducing the carboxylation efficiency of RuBisCO. This leads to an increase in the concentration of internal CO_2 in the substomatal chamber, leading to stomatal closure. This chapter discusses in detail, the ROS generation and the defense strategies adopted by the plants. It also throws light on the effect O_3 has on the physiological and biochemical processes operating in the plants.

References

Adir N, Zer H, Shochat I (2003) Photoinhibition – a historic perspective. Photosynth Res 76:343–370

Adrees M, Saleem F, Jabeen F, Rizwan M, Ali S, Khalid S, Ibrahim M, Iqbal N, Abbas F (2016) Effects of ambient gaseous pollutants on photosynthesis, growth, yield and grain quality of selected crops grown at different sites varying in pollution levels. Arch Agronom Sci 62(9):34–47

Agrawal M, Agrawal SB (1990) Effects of ozone exposure on enzymes and metabolitesof nitrogen-metabolism. Scientia Horticulture 43:169–177

Agrawal SB, Agrawal M (1999) Low temperature scanning electron microscope studies of stomatal response in snap bean plants treated with ozone and ethylenediurea. Biotronics 28:45–53

Agrawal GK, Rakwal R, Yonekura M, Saji H (2002) Rapid induction of defense/stress related proteins in leaves of rice (Oryza sativa) seedlings exposed to ozone is preceeded by newly phosphorylated proteins and changes in 66 K-Da ERK-type MAPK. J Plant Physiol 159:361–369

Ahsan N, Donnart T, Nouri MZ, Komatsu S (2010) Tissue-specific defense and thermo-adaptive mechanisms of soybean seedlings under heat stress revealed by proteomic approach. J Proteome Res 9:4189–4204

Altimir N, Kolari P, Tuovinen J-P, Vesala T, Back J, Suni T, Kulmala M, Hari P (2006) Foliage surface ozone deposition: a role for surface moisture? Biogeosciences 3:209–228

Andersen CP (2003) Source–sink balance and carbon allocation below ground in plants exposed to ozone. New Phytol 157:213–228

Antoni F (1994) Etude des interactions entre la pollution photo-oxydante (ozone), le stress hydrique et l'enrichissement en CO2 sur le métabolisme carboné chez l'épicéa et le hêtre. Master thesis, Université Henri Poincaré Nancy 1, France, pp 42

Apel K, Hirt H (2004) Reactive oxygen species: oxidative stress and signal transduction. Annu Rev Plant Biol 53:373–399

Ashraf M, Harris PJC (2013) Photosynthesis under stressful environments: An overview. Photosynthetica 51:163–190. https://doi.org/10.1007/s11099-013-0021-6

Athar H, Khan A, Ashraf M (2008) Exogenously applied ascorbic acid alleviates salt induced oxidative stress in wheat. Environ Exp Bot 63:224–231. https://doi.org/10.1016/j.envexpbot.2007.10.018

Bagard M, Le Thiec D, Delacôte E, Hasenfratz-Sauder M-P, Banvoy J, Ge'rard J, Dizengremel P, Jolivet Y (2008) Ozone-induced changes in photosynthesis and photorespiration of hybrid poplar in relation to the developmental stage of the leaves. Physiol Plant 134:559–574

Baier M, Kandlbinder A, Golidack D, Dietz K-J (2005) Oxidative stress and ozone: perception, signalling and response. Plant Cell Environ 28:1012–1020

Baishnab CT, Ralf O (2012) Reactive oxygen species generation and signaling in plants. Plant Signal Behav 7:1621–1633

Bandurska H, Borowaik K, Miara M (2009) Effect of two different ambient ozone concentrations on antioxidative enzymes in leaves of two tobacco cultivars with contrasting ozone sensitivity. Acta Biol Cracov Ser Bot 51(2):37–44

Baker NR, Rosenqvist E (2004) Applications of chlorophyll fluorescence can improve crop production strategies: an examination of future possibilities. J Exp Bot 55:1607–1621

Barcelo AR, Pomar F, Lopez-Serrano M, Pedreno MA (2003) Peroxidase: a multifunctional enzyme in grapevines. Funct Plant Biol 30:577–591

Bernacchi CJ, Leakey ADB, Heady LE, Morgan PB, Dohleman FG et al (2006) Hourly and seasonalvariation in photosynthesis and stomatal conductance of soybean grown at future CO2 and ozone concentrationsfor 3 years under fully open-air field conditions. Plant Cell Environ 29:20–90

Betz GA, Gerstner E, Stich S et al (2009) Ozone affects shikimate pathway genes and secondary metabolites in saplings of European beech (*Fagus sylvatica*L.) grown under greenhouse conditions. Trees 23:539–553

Betzelberger AM, Gillespie KM, McGrath JM, Koester RP, Nelson RL, Ainsworth EA (2010) Effects of chronic elevated ozone concentration on antioxidant capacity, photosynthesis and seed yield of 10 soybean cultivars. Plant Cell Environ 33:1569–1581

Betzelberger AM, Yendrek CR, Sun JD, Leisner CP, Nelson RL, Ort DR, Ainsworth EA (2012) Ozone exposure response for U.S. soybean cultivars: linear reductions in photosynthetic potential, biomass, and yield. Plant Physiol 160:1827–1839

Bhattacharjee S (2005) Reactive oxygen species and oxidative burst: roles in stress, senescence and signal transduction in plants. Curr Sci 89:58–67

Bhattacharjee S (2015) Membrane lipid peroxidation and its conflict of interest: the two faces of oxidative stress. Curr Sci 107(11):1811–1823

Biswas DK, Xu H, Li YG, Sun JZ, Wang XZ, Han XG, Jiang GM (2008) Genotypic differences in leaf biochemical, physiological and growth responses to ozone in 20 winter wheat cultivars released over the past 60 years. Glob Chang Biol 14:46–59

Biswas DK, Xu H, Li YG, Ma BL, Jiang GM (2013) Modification of photosynthesis and growth responses to elevated CO_2 by ozone in two cultivars of winter wheat with different years of release. J Exp Bot 64(6):1485–1496

Blokhina O, Virolainen E, Fagerstedt KV (2003) Antioxidants, oxidative damage and oxygen deprivation stress: A review. Ann Bot 91:179–194

Blokhina O, Fagerstedt KV (2010) Reactive oxygen species and nitric oxide in plant mitochondria: origin and redundant regulatory systems. Physiol Plant 138(4):447–462

Bolhar-Nordenkampf HR, Long SP, Baker NR, Oquist G, Schreiber U, Lechner EG (1989) Chlorophyll fluorescence as a probe of the photosynthetic competence of leaves in the field: a review of current instrumentation. Funct Ecol 3:497–514

Booker FL, Burkey KO, Jones AM (2012) Re-evaluating the role of ascorbic acid and phenolic glycosides in ozone scavenging in the leaf apoplast of *Arabidopsis thaliana* L. Plant Cell Environ 35(8):1456–1466

Burkey KO, Eason G, Fiscus EL (2003) Factors that affect leafextracellular ascorbic acid content and redox status. Physiol Plant 117:51–57

Burkey KO, Wei C, Eason G, Ghosh P, Fenner GP (2000) Antioxidant metabolite levels in ozone-sensitive and tolerant genotypes of snap bean. Physiol Plant 110:195–200

van Buuren ML, Guidi L, Fornalè S, Ghetti F, Franceschetti M, Soldatini GF, Bagni N (2002) Ozone-response mechanisms in tobacco: implications of polyamine metabolism. New Phytol 156:389–398

Cabané M, Pireaux J-C, Léger E, Weber E, Dizengremel P, Pollet B, Lapierre C (2004) Condensed lignins are synthesized in poplar leaves exposed to ozone. Plant Physiol 134:586–594

Calatayud A, Barreno E (2004) Response to ozone in two lettuce varieties on chlorophyll a fluorescence, photosynthetic pigments and lipid peroxidation. Plant Physiol Biochem 42:549–555

Calatayud A, Ramirez JW, Iglesias DJ, Barreno E (2002) Effects of ozone on photosynthetic CO_2 exchange, chlorophyll a fluorescence and antioxidant systems in lettuce leaves. Physiol Plant 116:308–316

Calatayud A, Iglesias D, Talon M, Barreno E (2003) Effects of 2 months ozone exposure in spinach leaves on photosynthesis, antioxidant systems and lipid peroxidation. Plant Physiol Biochem 41:839–845

Caregnato FF, Bortolin RF, Divan Junior AM, Moreira JCF (2013) Exposure to elevated ozone levels differentially affects the antioxidant capacity and the redox homeostasis of two subtropical Phaseolus vulgaris L. varieties. Chemosphere, 93(2):320–330

Caretto S, Linsalata V, Colella G, Mita G, Lattanzio V (2013) Carbon fluxes between primary metabolism and phenolic pathway in plant tissues under stress. Int J Mol Sci 16:26378–26394

Carlsson AS, Wallin G, Sandelius AS (1996) Species and age dependant sensitivity to ozone in young plants of pea, wheat and spinach: Effects on acyl lipid and pigment content and metabolism. Physiol Plant 98:271–280

Caregnato FF., Bortolin RF., Divan Junior AM., Moreira JCF. 2013. Exposure to elevated ozone levels differentially affects the antioxidant capacity and the redox homeostasis of two subtropical *Phaseolus vulgaris* L. varieties. Chemosphere, 93 (2): 320–330

Castagna A, Ranieri A (2009) Detoxification and repair process of ozone injury: from O uptake to gene expression adjustment. Environ Pollut 157:1461–1469

Castagna A, Nali C, Ciompi G, Lorenzini G, Soldatini GF, Ranieri A (2001) O₃ exposure effects photosynthesis of pumpkin (*Cucurbita pepo*) plants. New Phytol 152:223–229

Chaudhary N, Agrawal SB (2013) Intraspecific responses of six Indian clover cultivars under ambient and elevated levels of ozone. Environ Sci Pollut Res 20:5318–5329

Chaudhary N, Agrawal SB (2014) Role of gamma radiation in changing phytotoxic effect of elevated level of ozone in *Trifolium alexandrinum* L. (Clover). Atmos Pollu Res 5:104–112

Chaudhary N, Singh S, Agrawal SB, Agrawal M (2013) Assessment of six Indian cultivars of mung bean against ozone by using foliar injury index and changes in carbon assimilation, gas exchange, chlorophyll fluorescence and photosynthetic pigments. Environ Monit Assess 185:7793–7807

Chernikova T, Robinson JM, Lee EH, Mulchi CL (2000) Ozonetolerance and antioxidant enzyme activity in soybean cultivars. Photosynth Res 6:15–12

Cho K, Shibato J, Agrawal GK, Jung YH, Kubo A, Jwa NS, Tamogami S, Satoh K, Kikuchi S, Higashi T, Kimura S, Saji H, TanakaY IH, Masuo Y, Rakwal R (2008) Integrated transcriptomics, proteomics, and metabolomics analyses to survey ozone responses in the leaves of rice seedling. J Proteome Res 7:2980–2998

Cong T, Booker FL, Burkey KO, Hu S (2009) Elevated atmospheric carbon dioxide and O3 differentially alter nitrogen acquisition in peanut. Crop Sci 49:1827–1836

Conklin PL, Barth C (2004) Ascorbic acid, a familiar small molecule intertwined in the response of plants to ozone, pathogens and the onset of senescence. Plant Cell Environ 27:959–970

Conklin PL, Williams EH, Last RL (1996) Environmental stress sensitivity of an ascorbic acid-deficient *Arabidopsis* mutant. *Proc Natl Acad Sci U S A***93**:9970–9974

Dalstein L, Torti X, Le Thiec D, Dizengremel P (2002) Physiologicalstudy of declining Pinuscembra (L.) trees in southern France. Trees 16:299–305

Danielsson H, Karlsson PE, Pleijel H (2013) An ozone response relationship for four *Phleum pratense* genotypes based on modelling of the phytotoxic ozone dose (POD). Environ Exp Bot 90:70–77

Dann MS, Pell EJ (1989) Decline of activity and quantity of ribulose bisphosphate carboxylase oxygenase and net photosynthesis in ozone-treated potato foliage. Plant Physiol 91:427–432

Davison AW, Barnes JD (1998) Effects of ozone on wild plants. New Phytol 139:135e151

Del R'ıo LA, Sandalio LM, Corpas FJ, Palma JM, Barroso JB (2006) Reactive oxygen species and reactive nitrogen species in peroxisomes. Production, scavenging, and role in cell signaling. Plant Physiol 141(2):330–335

Degl'Innocenti E, Guidi L, Soldatini GF (2002) Characterization of the photosynthetic response of tobacco leaves to ozone: CO2 assimilation and chlorophyll fluorescence. J Plant Physiol 159:845–853

Deschaseaux A (1997) Effets de l'ozone sur les processus de fixation du CO2 chez Populus tremula x alba: approches enzymatique et isotopique (delta 13C). Master thesis, Université Henri Poincaré Nancy 1, France, pp 24

D'haese D, Vandermeiren K, Asard H, Horemans N (2005) Other factors than apoplastic ascorbate contribute to the differential ozone tolerance of two clones of Trifolium repens L. Plant Cell Environ 28:623–632

Di Baccio D, Castagna A, Paoletti E, Sebastiani L, Ranieri A (2008) Could the differences in O₃ sensitivity between two poplar clones be related to a difference in antioxidant defense and secondary metabolic response to O₃ influx? Tree Physiol 28:1761–1772

Dietz KJ (2016) Thiol-based peroxidases and ascorbate peroxidases: why plants rely on multiple peroxidase systems in the photosynthesizing chloroplast? Mol Cells 39:20–25

Dietz KJ, Turkan I, Krieger-Liszkay A (2016) Redox- and reactive oxygen species-dependent signaling in and from the photosynthesizing chloroplast. Plant Physiol 171:1541–1550

Dizengremel P (2001) Effects of ozone on the carbon metabolism of forest trees. Plant Physiol Biochem 39:729–742

Dizengremel P, Le Thiec D, Bagard M, Jolviet Y (2008) Ozone risk assessment for plants: Central role of metabolism dependant changes in reducing power. Environ Pollut 156:11–15

Drogoudi PD, Ashmore MR (2002) Effects of elevated ozone on yield and carbon allocation in strawberry cultivars differing in developmental stage. Phyton-Annales Rei Botanicae 42:45–53

Dumont J, Keski-Saari S, Keinänen M, Cohen D, Ningre N, KontunenSoppela S, Baldet P, Gibon Y, Dizengremel P, Vaultier M-N, Jolivet Y, Oksanen E, Le Thiec D (2014) Ozone affects ascorbate and glutathione biosynthesis as well as amino acid contents in three Euramerican poplar genotypes. Tree Physiol 34(3):253–266. https://doi.org/10.1093/treephys/tpu004.

Elstner EF (1991) Mechanism of oxygen activation in different compartments. In: Pell EJ, Steffen KL (eds) Active Oxygen/Oxidative Stress and Plant Metabolism. American Society of Plant Physiologists, Roseville, pp 13–25

Emberson LD, Ashmore MR, Cambridge HM, Simpson D, Tuovinen JP (2000) Modelling stomatal ozone flux across Europe. Environ Pollut 109:403–413

Ernst D, Jürgensen M, Bahnweg G, Heller W, Müller-Starck G (2012) Common links of molecular biology with biochemistry and physiology in plants under ozone and pathogen attack. In: Matyssek R, Schnyder H, Osswald W, Ernst D, Munch PH (eds) Growth and defence in plants—resource allocation at multiple scales, Ecological Studies, vol 220. Springer, Berlin, pp 29–52

Feng ZZ, Kobayashi K, Ainsworth EA (2008) Impact of elevated ozone concentration on growth, physiology and yield of wheat (*Triticum aestivum* L.): a meta-analysis. Glob Chang Biol 14:2696–2708

Feng ZZ, Pang J, Nouchi I, Kobayashi K, Yamakawa T, Zhu J (2010) Apoplastic ascorbate contributes to the differential ozone sensitivity in two varieties of winter wheat under fully open-air field conditions. Environ Pollut 158:3539–3545

Feng Z, Pang J, Kobayashi K, Zhu J, Ort DR (2011) Differential responses in two varieties of winter wheat to elevated ozone concentration under fully open-air field conditions. Glob Chang Biol 17:580–591

Feng Z, Sun J, Wan W, Hu E, Calatayud V (2014) Evidence of widespreadozone-induced visible injury on plants in Beijing, China. Environ Pollut 193:296–301

Feng Z, Wang L, Pleijel H, Zhu J, Kobayashi K (2016) Differential effects of ozone on photosynthesis of winter wheat among cultivars depend on antioxidative enzymes rather than stomatal conductance. Sci Total Environ 572:404–411

Feng ZZ, Kobayashi K, Wang XK, Feng ZW (2009) A meta-analysis of responses of wheat yield formation to elevated ozone concentration. Chin Sci Bull 54:249–255

Fiscus EL, Reid CD, Miller JE, Heagle AS (1997) Elevated CO_2 reduces O_3 flux and O_3 -induced yield losses in soybeans: possible implications for elevated CO2 studies. J Exp Bot 48:307–313

Flowers MD, Fiscus EL, Burkey KO, Booker FL, Dubois J-JB (2007) Photosynthesis, chlorophyll fluorescence, and yield of snap bean (*Phaseolus vulgaris* L.) genotypes differing in sensitivity to ozone. Environ Exp Bot 61:190–198

Fontaine V, Pelloux J, Podor M, Afif D, Gérant D, Grieu P, Dizengremel P (1999) Carbon fixation in Pinus halepensis submitted to ozone. Opposite response of ribulose-1,5-bisphosphate carboxylase/oxygenase and phosphoenolpyruvate carboxylase. Physiol Plant 105:187–192

Forde BG, Lea PJ (2007) Glutamate in plants: metabolism, regulation, and signaling. J Exp Bot 58:2339–2358

Foyer CH, Noctor G (2005) Oxidant and antioxidant signalling in plants: a re-evaluation of the concept of oxidative stress in a physiological context. Plant Cell Environ 28:1056–1071

Foyer CH, Noctor G (2000) Oxygen processing in photosynthesis: regulation and signaling. New Phytol 146(3):359–388

Foyer CH, Noctor G (2011) Ascorbate and glutathione: the heart of the redox hub. Plant Physiol 155:2–18

Foyer CH, Noctor G (2013) Redox signaling in plants. Antioxid Redox Signal 18:2087–2090. https://doi.org/10.1089/ars.2013.5278

Francini A, Nali C, Picchi V, Lorenzini G (2007) Metabolic changes in white clover exposed to ozone. Environ Exp Bot 60:11–19

Frei M, Wissuwa M, Pariasca-Tanaka J, Chen CP, Südekum KH, Kohno Y (2012) Leaf ascorbic acid level: is it really important for ozone tolerance in rice? Plant Physiol Biochem 59:63–70

Friend AL, Tomlinson PT (1992) Mild ozone exposure alters 14C dynamics in foliage of Pinus taeda L. Tree Physiol 11:215–227

Garg N, Manchanda G (2009) ROS generation in plants: boon or bane? Plant Biosyst 143:8–96

Gaucher C, Costanzo N, Afif D, Mauffette Y, Chevrier N, Dizengremael P (2003) The impact of elevated ozone and carbon di oxide on young *Acer saccharum* seedlings. Physiol Plant 117:392–402

Gerosa G, Finco A, Antonio Negri A, Marzuoli R, Wieser G (2013) Ozone Fluxes to a Larch Forest Ecosystem at the Timberline in the Italian Alps. https://doi.org/10.5772/56282

Giacomo B, Forino LMC, Tagliasacchi AM, Bernardi R, Durante M (2010) Ozone damage and tolerance in leaves of two poplar genotypes. Caryologia 63:422–434

Gill SS, Tuteja N (2010) Reactive oxygen species and antioxidant machinery in abiotic stress tolerance in crop plants. Plant Physiol Biochem 48:909–930

Gillespie KM, Rogers A, Ainsworth EA (2011) Growth at elevated ozone or elevated carbon dioxide concentration alters antioxidant capacity and response to acute oxidative stress in soybean (Glycine max). J Exp Bot 62:2667–2678

Gilroy S, Białasek M, Suzuki N, Górecka M, Devireddy AR, Karpiński S, Mittler R (2016) ROS, calcium, and electric signals: key mediators of rapid systemic signaling in plants. Plant Physiol 171(3):1606–1615

Glick RE, Schlagnhaufer CD, Arteca RN, Pell EJ (1995) Ozone-induced ethlylene emission accelerates the loss of ribulose-1, 5-bisphosphate carboxylase/oxygenase and nuclear-encoded mRNAs in senescing potato leaves. Plant Physiol 109:891–898

Godde D, Buchhold J (1992) Effect of long term fumigation with ozone on the turnover of the D-1 reaction centre polypeptide of photosystem II in spruce (*Picea abies*). Physiol Plant 86:568–574

Goumenaki E, Taybi T, Borland A, Barnes J (2010) Mechanisms underlying the impacts of ozone on photosynthetic performance. Environ Exp Bot 69(3):259–266

Grantz DA, Yang S (2000) Ozone impacts on allometry and root hydraulic conductance arenot mediated by source limitation not developmental age. J Exp Bot 51:919–927

Gratão PL, Polle A, Lea PJ, Azevedo RA (2005) Making the life of heavy metal-stressed plants a little easier. Funct Plant Biol 32:481–494

Gross GG (1977) Cell wall-bound malate dehydrogenase from horseradish. Phytochemistry 16(3):319–321

Guidi L, Degl'Innocenti E (2008) Ozone effects on high light induced photoinhibition in Phaseolus vulgaris. Plant Sci 174:590–596

Guidi L, Degl'Innocenti E, Martinelli F, Piras M (2009) Ozone effects on carbon metabolism in sensitive and insensitive Phaseolus cultivars. Environ Exp Bot 66:117–125

Guidi L, Di Cagno R, Soldatini GF (2000) Screening of beans cultivars for their response to ozone as evaluated by visible symptoms and leaf chlorophyll fluorescence. Environ Pollut 107:349–355

Guidi L, Nali C, Ciompi S, Lorenzini G, Soldatini GF (1997) The use of chlorophyll fluorescence and leaf gas exchange as methods for studying the different responses to ozone of two bean cultivars. J Exp Bot 48:173–179

Guidi L, Nali C, Lorenzini G, Filppi F, Soladatini GF (2001) Effect of chronic ozone fumigation on the photosynthetic process of poplar clones showing different sensitivity. Environ Pollut 113:245–254

Gunthardt-Goerg MS, McQuattie CJ, Maurer S, Frey B (2000) Visible and microscopy injury in leaves of five deciduous tree species related to current critical ozone levels. Environ Pollut 109:489–500

Guzy MR, Heath RL (1993) Response to ozone of varieties of common bean (*Phaseolus vulgaris* L). New Phytol 124:617–625

Hain FG (1987) Interaction of insects, trees and air pollutants. Tree Physiol 3:93–102

Hayes F, Mills G, Ashmore M (2009) Effects of ozone on inter- and intra-species competition and photosynthesis in mesocosms of *Lolium perenne* and *Trifolium repens*. Environ Pollut 157:208–214

Heath RL (2008) Modification of the biochemical pathways of plants induced by ozone: what are the varied routes to change? Environ Pollut 155:453–463

Heath RL, Taylor GE (1997) Physiological processes and plant responses to ozone exposure. In: Sanderman H, Welburn AR, Heath RL (eds) Forest decline and ozone. Springer, Berlin, p 317

Heiden AC, Hoffmann T, Kahl J, Kley D, Klockow D, Langebartels C, Mehlhorn H, Sandermann H, Schraudner M, Schuh G et al (1999) Emission of volatile organic compounds from ozone-exposed plants. Ecol Appl 9:1160–1167

Hewitt DKL, Mills G, Hayes F, Wilkinson S, Davies W (2014) Highlighting the threat from current and near-future ozone pollution to clover in pasture. Environ Pollut 189:111–117

Hewitt DKL, Mills G, Hayes F, Norris D, Coyle M, Wilkinson S, Davies W (2016) N-fixation in legumes: An assessment of the potential threat posed by ozone pollution. Environ Pollut 208:909–918

Holopainen JK, Kainulainen P, Oksanen J (1997) Growth and reproduction of aphids and levels of free amino acids in Scotspine and Norway spruce in an open-air fumigation with ozone. Glob Chang Biol 3:139–147

Hoshika Y, Carriero G, Feng Z, Zhang Y, Paoletti E (2014) Determinants of stomatal sluggishness in ozoneexposed deciduous tree species. Sci Total Environ 481:453–458

Hoshika Y, De Marco A, Materassi A, Paoletti E (2016) Light intensity affects ozone-induced stomatal sluggishness in snapbean. Water Air Soil Pollut 227:419–425

Hoshika Y, Watanabe M, Inada N, Koike T (2013) Model-based analysis of avoidance of ozone stress by stomatal closure in Siebold's beech (*Fagus crenata*). Ann Bot 112:1149–1158. https://doi.org/10.1093/aob/mct166. Available online at www.aob.oxfordjournals.org

Hu WH, Song XS, Shi K, Xia XJ, Zhou YH, JQ Y (2008) Changes in electron transport, superoxide dismutase and ascorbate peroxidase isoenzymes in chloroplasts and mitochondria of cucumber leaves as influenced by chilling. Photosynthetica 46(4):581–588

Huang S, Aken OV, Schwarzländer M, Belt K, Millar AH (2016) The roles of mitochondrial reactive oxygen species in cellular signaling and stress response in plants. Plant Physiol 171:1551–1559

Iglesias DJ, Calatayud A, Barreno E, Primo-Milloa E, Talon M (2006) Responses of citrus plants to ozone: leaf biochemistry, antioxidant mechanisms and lipid peroxidation. Plant Physiol Biochem 44:125–131

Imlay JA, Linn S (1988) DNA damage and oxygen radical toxicity. Science 240(4857):1302–1309. https://doi.org/10.1126/science.3287616

Inclan R, Gimeno BS, Dizengremel P, Sanchez M (2005) Compensation processes of Aleppo pine (*Pinus halepensis* Mill.) to ozone exposure and drought stress. Environ Pollut 137:517–524

IPCC (2013) Climate change 2013: the physical science basis. Contribution of working group I to the fifth assessment report of the intergovernmental panel on climate change. In: Stocker TF, Qin D, Plattner G-K, Tignor M, Allen SK, Boschung J, Nauels A, Xia Y, Bex V, Midgley PM (eds). Cambridge University Press, Cambridge, UK/New York, p 1535

Iriti M, Faoro F (2009) Chemical diversity and defence metabolism: how plants cope with pathogens and ozone pollution. Int J Mol Sci 10(8):3371–3399

Jajic I, Sarna T, Strzalka K (2015) Senescence, stress, and reactive oxygen species. Plants 4:393–411. https://doi.org/10.3390/plants4030393.

Jang SJ, Wi SJ, Choi YJ, An G, Park KY (2012) Increased polyamine biosynthesis enhances stress tolerance by preventing the accumulation of reactive oxygen species: T-DNA mutational analysis of *Oryza sativa* Lysine Decarboxylase-like protein. Mol Cells 34(3):251–262

Joo JH, Wang S, Chen JG, Jones AM, Fedoroff N (2005) Different signaling and cell death roles of heterotrimeric G protein a and b subunits in the *Arabidopsis* oxidative stress response to ozone. Plant Cell 17:957–970

Jones CG, Hartley SE (1999) A protein competition model of phenolic allocation. Oikos 86:27–44. https://doi.org/10.2307/3546567

Jud W, Fischer L, Canaval E, Wohlfahrt G, Tissier A, Hansel A (2016) Plant surface reactions: an opportunistic ozone defence mechanism impacting atmospheric chemistry. Atmos Chem Phys 16:277–292

Kainulainen P, Holopainen JK, Holopainen T (2000) Combined effects of ozone and nitrogen on secondary compounds, amino acids, and aphid performance in Scots pine. J Environ Qual 29:334–342

Kaiser WM (1979) Reversible inhibition of the Calvin cycle and activation of oxidative pentose phosphate cycle in isolated intact chloroplasts by hydrogen peroxide. Planta 145(4):377–382

Kaiser WM, Huber SC (2001) Post-translational regulation of nitratereductase: mechanism, physiological relevance and environmental triggers. J Exp Bot 52:1981–1989

Kangasjarvi J, Jaspers P, Kollist H (2005) Signalling and cell death in ozone-exposed plants. Plant Cell Environ 28:1021–1036

Karlsson PE, Braun S, Broadmeadow M, Elvira S, Emberson L, Gimeno BS, Le Thiec D, Novak K, Oksanen E, Schaub M, Uddling J, Wilkinson M (2007) Risk assessments for forest trees: The performance of the ozone flux versus the AOT concepts. Environ Pollut 146:608–616

Karuppanapandian T, Moon J-C, Kim C, Manoharan K, Kim W (2011) Reactive oxygen species in plants: their generation, signal transduction, and scavenging mechanisms. Aust J Crop Sci 5(6):709–725

Kaur G, Sharma A, Guruprasad K, Pati PK (2014) Versatile roles of plant NADPH oxidases and emerging concepts. Biotechnol Adv 32:551–563

Kerchev PI, Waszczak C, Lewandowska A et al (2016) Lack of GLYCOLATE OXIDASE 1, but not GLYCOLATE OXIDASE 2, attenuates the photorespiratory phenotype of CATALASE2-deficient Arabidopsis. Plant Physiol 171:1704–1719

Khan S, Andralojc PJ, Lea PJ, Parry MA (1999) 2'-carboxy-D-arabinitol 1-phosphate protects ribulose 1, 5-bisphosphate carboxylase/oxygenase against proteolytic breakdown. Eur J Biochem 266(3):840–847

Kitajima H, Butler WL (1975) Quenching of chlorophyll fluorescence and primary photochemistry in chloroplasts by dibromothynoquinone. Biochim Biophys Acta 376:105–115

Koch JR, Scherzer AJ, Eshita SM, Davis KR (1998) Ozonesensitivity in hybrid poplar is correlated with a lack of defense-gene activation. Plant Physiol 118:1243–1252

Kollner B, Krause GHM (2003) Effects of two different ozone exposure regimes on chlorophyll and sucrose content of leaves and yield parameters of sugar beet (*Beta vulgaris* L.) and rape (*Brassica napus* L.) Water Air Soil Pollut 144:317–332

Krause GH, Weis E (1991) Chlorophyll fluorescence and photosynthesis: the basics. Annu Rev Plant Physiol Plant Mol Biol 42:313–349

Lambers H, Chapin FS III, Pons TL (2008) Plant physiological ecology, 2nd-edition edn. Springer-Verlag, New York, p 640

Langebartels C, Kerner K, Leonardi S, Schraudner M, Trost M, Heller W, Sandermann H Jr (1991) Biochemical plant responses to ozone 1. Differential induction of polyamine and ethylene biosynthesis in tobacco. Plant Physiol 95:882–889

Langebartels C, Wohlgemuth H, Kschieschan S, Grun S, Sandermann H (2002) Oxidative burst and cell death in ozone exposed plants. Plant Physiol Biochem 40:567–575

Larcher W (2003) Physiological plant ecology, 4th edn.Springer,Berlin.Forde BG, Lea PJ. 2007. Glutamate in plants: metabolism, regulation and signaling. J Exp Bot 58:2339–2358

Liu X, Sui L, Huang Y, Geng C, Yin B (2015) Physiological and visible injury responses in different growth stages of winter wheat to ozone stress and the protection of spermidine. Atmos Pollut Res 6:596–604

Loreto F, Fares S (2007) Is ozone flux inside leaves only a damage indicator? Clues from volatile isoprenoid studies. Plant Physiol 143:1096–1100. https://doi.org/10.1104/pp.106.091892.

Loreto F, Pinelli P, Manes F, Kollist H (2004) Impact of ozone on monoterpene emissions and evidence for an isoprene-like antioxidant action of monoterpenes emitted by *Quercus ilex* leaves. Tree Physiol 24(4):361–367. https://doi.org/10.1093/treephys/24.4.361.

Lutz C, Anegg S, Gerant D, Alaoui-Sosse B, Gerard J, Dizengremel P (2000) Beech trees expose to high CO2 and to simulated summer ozone levels: effects on photosynthesis, chloroplast components and leaf enzyme activity. Physiol Plant 109:252–259

Maccarrone M, Veldink GA, Vliegenthart FG, Finazzi Agro A (1997) Ozone stress modulates amine oxidase and lipoxygenase expression in (*Lens culinaris*) seedlings. FEBS Lett 408:241–244

Maddison J, Lyons T, Plochl M, Barnes J (2002) Hydroponically cultivated radish fed l-galactono-1, 4-lactone exhibit increased tolerance to ozone. Planta 214:383–391

Maier-Maercker U (1999) Predisposition of trees to drought stress by ozone. Tree Physiol 19:71–78

Manderscheid R, Jager HJ, Kress LW (1992) Effects of ozone on foliar nitrogen metabolism of *Pinus taeda* L. and implications for carbohydrate metabolism. New Phytol 121:623–633

Marino D, Dunand C, Puppo A, Pauly N (2012) A burst of plant NADPH oxidases. Trends Plant Sci 17(1):9–15. https://doi.org/10.1016/j.tplants.2011

Marre MT, Amicucci E, Zingarelli L, Albergoni F, Marre E (1998) The respiratory burst and electrolyte leakage induced by sulfhydryl blockers in *Egeria densa* leaves are associated with H_2O_2 production and are dependent on Ca^{2+} influx. Plant Physiol 118:1379–1387

McAinsh MR, Evans NH, Montgomery LT, North KA (2002) Calcium signalling in stomatal responsesto pollutants. New Phytol 153:441–447

Mckee IF, Bullimore JF, Long SP (1997) Will elevated CO protect the yield of wheat from O damage? Plant Cell Environ 20:77–84

Mignolet-Spruyt L, Xu E, Idänheimo N, Hoeberichts FA, Mühlenbock P, Brosché M, Van Breusegem F, Kangasjärvi J (2016) Spreading the news: subcellular and organellar reactive oxygen species production and signalling. J Exp Bot 67(13):3831–3844

Miller G, Suzuki N, Ciftci-Yilmaz S, Mittler R (2010) Reactive oxygen species homeostasis and signaling during drought and salinity stresses. Plant Cell Environ 33:453–467. https://doi.org/10.1111/j.1365-3040.2009.02041.x

Mills G, Buse A, Gimeno B, Bermejo V, Holland M, Emberson L, Pleijel H (2007) A synthesis of AOT40-based response functions and critical levels of ozone for agricultural and horticultural crops. Atmos Environ 41:2630–2643

Mills G, Pleijel H, Buker P et al (2010) Mapping critical levels for vegetation. Revision undertaken in Summer 2010 to include new flux- based critical levels and response functions for ozone, in: Mapping Manual 2004. International Cooperative Programme on Effects of Air Pollution on Natural Vegetation and Crops. http://icpvegetation.ceh.ac.uk/manuals/mapping_manual.html

Mills G, Pleijel H, Braun S, Büker P, Bermejo V, Calvo E, Danielsson H, Emberson L, Fernandez IG, Grünhage L, Harmens H, Hayes F, Karlsson PE, Simpson D (2011) New stomatal flux-based critical levels for ozone effects on vegetation. Atmos Environ 45(28):5064–5068. https://doi.org/10.1016/j.atmosenv.2011.06.009

Mishra AK, Agrawal SB (2015) Biochemical and physiological characteristics of tropical mung bean (*Vigna radiata* L.) cultivars against chronic ozone stress: an insight to cultivar-specific response. Protoplasma 252:797–811

Mishra AK, Rai R, Agrawal SB (2013) Individual and interactive effects of elevated carbon dioxide and ozone on tropical wheat (*Triticum aestivum* L.) cultivars with special emphasis on ROS generation and activation of antioxidant defense system. Indian J Biochem Biophys 50:139–149

Mittler R (2002) Oxidative stress, antioxidants and stress tolerance. Trends Plant Sci 7:405–410

Mittler R, Vanderauwera S, Gollery M, Van Breusegem F (2004) Reactive oxygen gene network of plants. Trends Plant Sci 9:490–498

Møller IM, Kristensen BK (2004) Protein oxidation in plant mitochondria as a stress indicator. Photochem Photobiol Sci 3(8):730–735

Moller IM, Jensen PE, Hansson A (2007) Oxidative modifications to cellular components in plants. Annu Rev Plant Biol 58:459–481

Morgan PB, Ainsworth EA, Long SP (2003) How does elevated ozone impact soybean? A meta analysis of photosynthesis, growth and yield. Plant Cell Environ 26:1317–1328

Mullineaux PM, Baker NR (2010) Oxidative stress: antagonistic signaling for acclimation or cell death? Plant Physiol 154:521–525

Murchie EH, Lawson T (2013) Chlorophyll fluorescence analysis: a guide to good practice and understanding some new applications. J Exp Bot 64(13):3983–3998

Nakabayashi R, Saito K (2015) Integrated metabolomics for abiotic stress responses in plants. Curr Opin Plant Biol 24:10–16

Niu J, Feng Z, Zhang W, Zhao P, Wang X (2014) Non-stomatal limitation to photosynthesis in Cinnamomum camphora seedings exposed to elevated O3. PLoS One 9(6):e98572. https://doi.org/10.1371/journal.pone.0098572

Noctor G, Foyer CH (1998) Ascorbate and glutathione: keeping active oxygen under control. Annu Rev Plant Physiol Plant Mol Biol 49:249–279

Noctor G, Mhamdi A, Chaouch S, Han YI, Neukermans J, Marquez-Garcia B et al (2012) Glutathione in plants: an integrated overview. Plant Cell Environ 35:454–484. https://doi.org/10.1111/j.1365-3040.2011.02400.x

Overmyer K, Tuominen H, Kettunen R, Betz C, Langebartels C, Sandermann H Jr, Kangasjarvi J (2000) The ozone sensitive *Arabidopsis rcd1* mutant reveals opposite roles for ethylene and jasmonate signaling pathways in regulating superoxide-dependent cell death. Plant Cell 12:1849–1862

Orvar BL, Ellis BE (1997) Transgenic tobacco plants expressing antisense RNA for cytosolic ascorbate peroxidase show increased susceptibility to ozone injury. Plant J 11(6):1297–1305

Pang J, Kobayashi K, Zhu J (2009) Yield and photosynthetic characteristics of flag leaves in Chinese rice (Oryza sativa L.) varieties subjected to free-air release of ozone. Agric Ecosyst Environ 132:203–211

Paoletti E, Grulke NE (2005) Does living in elevated CO2 ameliorate tree responseto ozone? A review on stomatal responses. Environ Pollut 137:483–493

Paoletti E, Grulke NE (2010) Ozone exposure and stomatal sluggishness in different plant physiognomic classes. Environ Pollut 158:2664–2671

Paolacci AR, D'ovidio R, Marabottini R, Nali LG, Abanavoli MR, Badiani M (2001) Ozone induces adifferential accumulation of phenylalanine ammonia-lyase, chalcone synthase and chalcone isomerase RNAtranscripts in sensitive and resistant bean cultivars. Aust J Plant Physiol 28:425–428

Paoletti E, Contran N, Bernasconi P, Günthardt-Goerg MS, Vollenweider P (2009) Structural and physiological responses to ozone in Manna ash (*Fraxinus ornus* L.) leaves of seedlings and mature trees under controlled and ambient conditions. Sci Total Environ 407:1631e1643. https://doi.org/10.1016/j.scitotenv.2008.11.061

Pasqualini S, Batini P, Ederli L, Antonielli M (2001) Effects of short-term ozone fumigation on tobacco plants: Response of the scavenging system and expression of the glutathione reductase. Plant Cell Environ 24(2):245–252

Pellegrini E (2014) PSII photochemistry is the primary target of oxidative stressimposed by ozone in *Tilia americana*. Urban For Urban Green 13:94–102

Pelloux J, Jolivet Y, Fontaine V, Banvoy J, Dizengremel P (2001) Changes in Rubisco and Rubisco activase gene expression and polypeptide content in Pinus halepensis M. Plant Cell Environ 24:123–131

Pellegrini E, Francini A, Lorenzini G, Nali C (2011) PSII photochemistry andcarboxylation efficiency in Liriodendron tulipifera under ozone exposure. Environ Exp Bot 70:217–226

Pesaresi P, Hertle A, Pribil M et al (2009) Arabidopsis STN7 kinase provides a link between short- and long-term photosynthetic acclimation. Plant Cell 21:2402–2423

Petrov VD, VanBreusegem F (2012) Hydrogen peroxide- a central hub for information flow in plant cells. AoBPlants. pls014. https://doi.org/10.1093/aob-pla/pls014

Pino ME, Mudd JB, Bailey-Serres J (1995) Ozone-induced alterations in the accumulation of newly synthesized proteins in leaves of maize. Plant Physiol 108:777–785

Possell M, Loreto F (2013) The role of volatile organic compounds in plant resistance to abiotic stresses: responses and mechanisms. In: Niinemets Ü, Monson RK (eds) Biology, controls and models of tree volatile organic compound emissions, vol 5. Springer, Dordrecht, pp 209–235

Quan LJ, Zhang B, Shi WW, Li HY (2008) Hydrogen peroxide in plants: a versatile molecule of the reactive oxygen species network. J Integr Plant Biol 50:2–18

Quinlan CL, Perevoshchikova IV, Hey-Mogensen M, Orr AL, Brand MD (2013) Sites of reactive oxygen species generation by mitochondria oxidizing different substrates. Redox Biol 1:304–312

Rai R, Agrawal M (2008) Evaluation of physiological and biochemical responses of two rice (Oryza sativa L.) cultivars to ambient air pollution using open top chambers at rural site in India. Sci Total Environ 407:679–691

Rai R, Agrawal M, Agrawal SB (2007) Assessment of yield losses in tropical wheat using open top chambers. Atmos Environ 41:9543–9554

Rai R, Agrawal M, Agrawal SB (2011) Effects of ambient O on wheat during reproductivedevelopment: gas exchange, photosynthetic pigments, chlorophyll fluorescence and carbohydrates. Photosynthetica 49:285–294

Rai R, Agrawal M (2014) Assessment of competitive ability of two Indian wheat cultivars under ambient O at different developmental stages. Environ Sci Pollut Res 21:1039–1053

Ranieri A, D'Ilrso G, Nali C, Lorenzini G, Soldatini GF (1996) Ozone stimulates apoplastic antioxidant systems in pumpkin leaves. Physiol Plant 97:381–387

Ranieri A, Castagna A, Soldatini GF (2000) Differential stimulation of ascorbate peroxidase isoforms by ozone exposure in sunflower plants. J Plant Physiol 156:266–271

Rainieri A, Giuntini D, Ferraro F, Nali B, Baldan G, Lorenzini G, Soldatini GF (2001) Chronic ozone fumigation induces alterations in thylakoid functionality and composition in two poplar clones. Plant Physiol Biochem 39:999–1008

Rao MV, Paliyath G, Ormrod DP (1995) Ultraviolet B and ozone induced biochemical changes in antioxidant enzymes of Arabidopsis thaliana. Plant Physiol 109:421–432

Rao MV, Davis KR (1999) Ozone-induced cell death occurs via two distinct mechanisms in Arabidopsis: the role of salicylic acid. Plant J 17:603–614

Rees DC, Tezcan FA, Haynes CA, Walton MY, Andrade S, Einsle O, Howard JB (2005) Structural basis of biological nitrogen fixation. Philos Trans R Soc A 363:971–984

Reichenauer T, Bolhar- Nordenkempf HR, Ehrlich U, Soja G, Postl WF, Halbwachs F (1997) The influence of ambient and elevated O_3 concentrations on photosynthesis in Populus nigra. Plant Cell Environ 20:1061–1069

Reiling K, Davison AW (1992) The response of native, herbaceous species toozone: growth and fluorescence screening. New Phytol 120:29–37

Renaut J, Bohler S, Hausman JF, Hoffmann L, Sergeant K, Ahsan N, Jolivet Y, Dizengremel P (2009) The impact of atmospheric composition on plants: a case study of ozone and poplar. Mass Spectrom Rev 28:495–516

Richter C (1992) Reactive oxygen and DNA damage in mitochondria. Mutat Res 275(3–6):249–255

Richards BL, Middleton JT, Hewitt WB (1958) Air pollution with reference to agronomic crops. Agron J 50:559–561

Roach T, Krieger-Liszkay A (2014) Regulation of Photosynthetic Electron Transport and Photoinhibition. Curr Protein Pept Sci 15:351–362

Robinson D (2005) Integrated root responses to variations in nutrient supply. In: BassiriRad (ed) Nutrient acquisition by plants. Springer, Berlin, pp 43–61

Rouhier N, Gelhaye E, Jacquot JP (2004) Plant glutaredoxin: still mysterious reducing systems. Cell Mol Life Sci 61:1266–1277

Sagi M, Fluhr R (2006) Production of reactive oxygen species by plant NADPH oxidases. Plant Physiol 141:336–340

Sakaki T, Kondo N, Yamada M (1990a) Pathway for the synthesis of triacylglycerol from mono-galactosyldiacylglycerols in ozone-fumigated spinach leaves. Plant Physiol 94:773–780

Sakaki T, Kondo N, Yamada M (1990b) Free fatty acids regulate two galactosyltransferases in chloroplast envelope mambranes isolated from spinach leaves. Plant Physiol 94:781–787

Sakaki T, Saito K, Kawaguchi A, Kondo N, Yamada M (1990c) Conversion of monogalacto-syldiacylglycerols to triacylglycerol in ozone-fumigated spinach leaves. Plant Physiol 94:766–772

Sakaki T, Kondo N, Sugahara K (2008) Breakdown of photosynthetic pigments and lipids in spinach leaves with ozone fumigation: role of active oxygens. Physiol Plant 59:28–34

Sandermann H, Ernst D, Heller W, Langebartels C (1998) Ozone: an abiotic elicitor of plant defence reactions. Trends Plant Sci 3:47–50

Sanmartin M, Drogoudi PD, Lyons T, Pateraki I, Barnes J, Kanellis AK (2003) Over-expression of ascorbate oxidase in the apoplast of transgenic tobacco results in altered ascorbate and glutathione redox states and increased sensitivity to ozone. Planta 216:918–928

Sarkar A, Rakwal R, Agrawal SB, Shibato J, Ogawa Y, Yoshida Y, Agrawal GK, Agrawal M (2010) Investigating the impact of elevated levels of ozone on tropical wheat using integrated phenotypical, physiological, biochemical and proteomics approaches. J Proteome Res 9:4565–4584

Sarkar A, Agrawal SB (2010) Identification of ozone stress in Indian rice through foliar injury and differential protein profile. Environ Monit Assess 161:205–215

Sarkar A, Singh AA, Agrawal SB, Ahmed A, Rai SP (2015) Cultivar specific variations in anti-oxidative defense system, genome and proteome of two tropical rice cultivars against ambi-ent and elevated ozone. Ecotoxicol Environ Saf 115:101–111

Scebba F, Canaccini F, Castagna A, Bender J, WeigelHJ RA (2006) Physiological and biochemi-cal stress responses in grassland species are influenced by both early-season ozone exposure and interspecific competition. Environ Pollut 142:540–548

Schieber M, Chandel NS (2015) ROS function in redox signaling and oxidative stress. Curr Biol 24(10):R453–R462

Sgarbi E, Fornasiero RB, Lins AP, Bonatti PM (2003) Phenol metabolism is differentially affected by ozone in two cell lines from grape (Vitis vinifera L.) leaf. Plant Sci 165:951–957

Sharma P, Jha AB, Dubey RS, Pessarakli M (2012) Reactive oxygen species, oxidative damage, and antioxidative defense mechanism in plants under stressful conditions. J Bot 2012:1–26

Shokolenko IN, Wilson GL, Alexeyev MF (2014) Aging: a mitochondrial DNA perspective, critical analysis and an update. World J Exp Med 4(4):46–57

Singh E, Tiwari S, Agrawal M (2009) Effects of elevated ozone on photosynthesis and stomatal conductance of two soybean varieties: a case study to assess impacts of one component of predicted global climate change. Plant Biol 11(Suppl. 1):101–108

Singh E, Tiwari S, Agrawal M (2010) Variability in antioxidant and metabolite levels, growth and yield of two soybean varieties: an assessment of anticipated yield losses under projected elevation of ozone. Agric Ecosyst Environ 135(3):168–177

Singh AA, Agrawal SB, Shahi JP, Agrawal M (2014) Assessment of growth and yield losses in two Zea mays L. cultivars (quality protein maize and nonquality protein maize) under pro-jected levels of ozone. Environ Sci Pollut Res 21:2628–2641

Sitch S, Cox PM, Collins WJ, Huntingford C (2007) Indirect radiative forcing of climate change through ozone effects on the land-carbon sink. Nature 448(7155):791–794

Sitch S., Cox PM., Collins WJ., Huntingford C. 2007. Indirect radiative forcing of climate change through ozone effects on the land-carbon sink. Nature Letters. doi:10.1038/nature06059.

Smith H, Neyra C, Brennan E (1990) The relationship between foliar injury, nitrogen metabolism, and growth parameters in ozonated soybeans. Environ Pollut 63(79):93

Sprent J (2007) Evolving ideas of legume evolution and diversity; a taxonomic perspective on the occurrence of nodulation. New Phytol 174:11–25

Srinandhinidevi KM, Khopade M, Pangsatabam KD (2015) A preliminary study on the effects of ozone on induction of resistance in *Cicer arietinum* and *Trigonella foenum* against acute ozone exposure. IOSR J Biotechnol Biochem 1(5):06–14

Sun JD, Feng ZZ, Ort DR (2014) Impacts of rising tropospheric ozone on photosynthesis and metabolite levels on field grown soybean. Plant Sci 226:147–161

Takahashi M, Asada K (1988) Superoxide production in aprotic interior of chloroplast thylakoids. Arch Biochem Biophys 267:714–722

Takahama U, Oniki T (1997) A peroxidase/phenolics/ascorbate system can scavenge hydrogen peroxide in plant cells. Physiol Plant 101:845–852

Tamaoki M, Matsuyama T, Kanna M, Nakajima N, Kubo A, Aono M, Saji H (2003) Differential ozone sensitivity among *Arabidopsis* accessions and its relevance to ethylene synthesis. Planta 216:552–560

Tausz M, Grulke NE, Wieser G (2007) Defense and avoidance of ozone under global change. Environ Pollut 147:525–531

Temple MD, Perrone GG, Dawes IW (2005) Complex cellular responses to reactive oxygen species. Trends Cell Biol 15:319–326

Tetteh R, Yamaguchi M, Wada Y, Izuta T (2015) Effects of ozone on growth, net photosynthesis and yield of two African varieties of Vigna unguiculata. Environ Pollut 196:230–238

Tian S, Wang X, Li P, Wang H, Ji H, Xie J, Qiu Q, Shen D, Dong H (2016) Plant aquaporin AtPIP1; 4 links apoplastic H2O2 induction to disease immunity pathways. Plant Physiol 171:1635–1650

Tiwari S, Agrawal M, Marshall FM (2006) Evaluation of ambient air pollution impact on carrot plants at a suburban site using open top chambers. Environ Monit Assess 119:15–30

Tiwari S, Agrawal M (2011) Assessment of the variability in response of radish and brinjal at biochemical and physiological levels under similar ozone exposure conditions. Environ Monit Assess 175(1–4):443–454

Tiwari S, Agrawal M (2009) Protection of palak (Beta vulgaris L. var Allgreen) plants from ozone injury by ethylenediurea (EDU): Roles of biochemical and physiological variations in alleviating the adverse impacts. Chemosphere 75:1492–1499

Torsethaugen G, Pell EJ, Assmann SM (1999) Ozone inhibits guard cell K+ channels implicated in stomatal opening. Proc Natl Acad Sci U S A 96:13577–13582

Toderova D, Katerova Z, Alexieva V, Sergiev I (2015) Polyamines – possibilities for application to increase plant tolerance and adaptation capacity to stress. Genet Plant Physiol 5(2):123–144

Torsethaugen G, Pitcher LH, Zilinskas BA, Pell EJ (1997) Overproduction of ascorbate peroxidase in the tobacco chloroplast does not provide protection against ozone. Plant Physiol 114:529–537

Tripathy BC, Oelmüller R (2012) Reactive oxygen species generation and signaling in plants. Plant Signal Behav 712:1621–1633

Tripathi R, Agrawal SB (2013) Interactive effect of supplemental ultraviolet B and elevated ozone on seed yield and oil quality of two cultivars of linseed (*Linum usitatissimum* L.) carried out in open top chambers. J Sci Food Agr 93:1016–1025

Udding J, Karlsson PE, Glorvigen A, Sellden G (2005) Ozone impairs autumnal resorption of nitrogen from birch (Betulapendula) leaves, causing an increase in whole-tree nitrogen loss through litter fall. Tree Physiol 26:113–120

Udvardi M, Poole PS (2013) Transport and metabolism in legume-rhizobia symbiosis. Annu Rev Plant Biol 64:781–805

Ueda Y, Uehara N, Sasaki H, Kobayashi K, Yamakawa T (2013) Impacts of acute ozone stress on superoxide dismutase (SOD) expression and reactive oxygen species (ROS) formation in rice leaves. Plant Physiol Biochem 70:396–402

Umponstira C, Kawayaskul S, Chuchaung S, Homhaul W (2009) Effect of Ozone on Nitrogen Fixation, Nitrogenase Activity and Rhizobium of Cowpea (Vigna unguiculata (L.) Walp). Naresuan University Journal 17(3):213–220

UNECE (2004) Revised manual on methodologies and criteria for mapping critical levels/loads and geographical areas where they are exceeded. www.icpmapping.org (February 12, 2006)

Vaahtera L, Brosché M, Wrzaczek M, Kangasjärvi J (2014) Specificity in ROS signaling and transcript signatures. Antioxid Redox Signal 21(9):1422–1441

Vahala J, Schlagnhaufer CD, Pell EJ (1998) Induction of an ACC synthase cDNA by ozone in light-grown *Arabidopsis thaliana* leaves. Physiol Plant 103:45–50

Vahisalu T, Puzorjoa I, Brosche M, Valk E, Lepiku M et al (2010) Ozone-triggered rapid stomatalresponse involves the production of reactive oxygen species and is controlled by SLAC1 and OST1. Plant J 162(3):442–453

Velikova V, Sharkey TD, Loreto F (2012) Stabilization of thylakoid membranes in isoprene-emitting plants reduces formation of reactive oxygen species. Plant Signal Behav 7(1):139–141. https://doi.org/10.4161/psb.7.1.18521

Wagg S, Mills G, Hayes F, Wilkinson S, Davies WJ (2013) Stomata are less responsive to environmental stimuli in high background ozone in *Dactylis glomerata* and *Ranunculus acris*. Environ Pollut 175:82–91. https://doi.org/10.1016/j.envpol.2012.11.027. PMID: 23354156

Wahid A (2006a) Influence of atmospheric pollutants on agriculture in developing countries: a case study with three new wheat varieties in Pakistan. Sci Total Environ 371:304–313

Wahid A (2006b) Productivity losses in barley attributable to ambient atmospheric pollutants in Pakistan. Atmos Environ 40:5342–5354

Wang J, Zeng Q, Zhu J, Tang H (2013) Dissimilarity of ascorbate glutathione (AsA-GSH) cycle mechanism in two rice (Oryza sativa L.) cultivars under experimental free air ozone exposure. Agric Ecosyst Environ 165:39–49

Wang J, Zeng Q, Zhu J, Chen C, Liu G, Tang H (2014) Apoplastic antioxidant enzyme responses to chronic free-air ozone exposure in two different ozone-sensitive wheat cultivars. Plant Physiol Biochem 82:183–193

Warren CR, Dreyer E, Adams MA (2003) Photosynthesis-Rubisco relationships in foliage of Pinus sylvestris in response to nitrogen supply and the proposed role of Rubisco and amino acids as nitrogen stores. Trees-Structure and Function 17:359–366

Watanabe M, Yamaguchi M, Tabe C, Iwasaki M, Yamashita R, Funada R, Fukami M, Matsumura H, Kohno Y, Izuta T (2007) Influences of nitrogen load on the growth and photosynthetic responses of *Quercus serrata* seedlings to O_3. Trees 21:421–432

Watanabe M, Hoshika Y, Koike T (2014) Photosynthetic responses of Monarch birch seedlings to differing timings of free air ozone fumigation. J Plant Res 127(2):339–345

Whitaker BD, Lee EH, Rowland RA (1990) Ethylenediurea and ozone protection: foliar glycerolipids and steryl lipids in snap bean exposed to ozone. Physiol Plant 80:286–293

Wilkinson S, Mills G, Illidge R, Davies WJ (2011) How is ozone pollution reducing our food supply? J Exp Bot 63:527–536

Wittig VE, Ainsworth EA, Long SP (2007) To what extent do current and projected increases in surface ozone affect photosynthesis and stomatal conductance of trees? A meta-analytic review of the last 3 decades of experiments. Plant Cell Environ 30(9):1150–1162

Wittig VE, Ainsworth EA, Naidu SL, Karnosky DF, Long SP (2009) Quantifying the impact of current and future tropospheric ozone on tree biomass, growth, physiology and biochemistry:a quantitative meta-analysis. Global Chang Biology 15:396–424

Yadav DK, Prasad A, Kruk J, Pospíšil P (2014) Evidence for the involvement of loosely bound plastosemiquinones in superoxide anion radical production in photosystem II. PLoS One 9:e115466. https://doi.org/10.1371/journal.pone.0115466

Yamaguchi M, Watanabe M, Iwasaki M, Tabe C, Matsumura H, Kohno Y, Izuta T (2007a) Growth and photosynthetic responses of Fagus crenata seedlings to O3 under different nitrogen loads. Trees 21:707–718

Yamaguchi M, Watanabe M, Matsuo N, Naba J, Funada R, Fukami M, Matsumura H, Kohno Y, Izuta T (2007b) Effects of nitrogen supply on the sensitivity to O3 of growth and photosynthesis of Japanese beech (*Fagus crenata*) seedlings. Water Air Soil Pollut Focus 7:131–136

Yamaguchi M, Watanabe M, Matsumura H, Kohno Y, Izuta T (2010) Effects of ozone on nitrogen metabolism in the leaves of *Fagus crenata* seedlings under different soil nitrogen loads. Trees 24:175–184. https://doi.org/10.1007/s00468-009-0391-3

Yamamoto HY, Akasada T (1995) Degradation of antenna chlorophyll binding protein CP43 during photoinhibition of PS II. Biochemistry 28:9038–9045

Yamaguchi M, Watanabe M, Matsumura H, Kohno Y, Izuta T (2010) Effects of ozone on nitrogen metabolism in the leaves of Fagus crenata seedlings under different soil nitrogen loads. Trees 24:175–184

Yendrek CR, Ainsworth EA (2015) A comparative analysis of transcriptomic, biochemical, and physiological responses to elevated ozone identifies species-specific mechanisms of resilience in legume crops. J Exp Bot 66(22):7101–7112

Zhang J, Schaub M, Ferdinand JA, Skelly JM, Steiner KC, Savage JE (2010) Leaf age affects the responses of foliar injury and gas exchange to tropospheric ozone in *Prunus serotina* seedlings. Environ Pollut 158:2627–2634

Zhang WW, Niu J, Wang X, Feng Z (2011) Effects of ozone exposure on growth and photosynthesis of the seedlings of Liriodendron chinense (Hemsl.) Sarg, a native tree species of subtropical China. Photosynthetica 49(1):29–36

Zhang W, Wang G, Liu X, Feng Z (2014) Effects of O3 exposure on seed yield, N concentration and photosynthesis of nine soybean cultivars (*Glycine max* (L.) Merr.) in Northeast China. Plant Sci 226:172–181

Zhao F, Song C-P, He J, Zhu H (2007) Polyamines Improve K+/Na+ Homeostasis in Barley Seedlings by Regulating Root Ion Channel Activities. Plant Physiol 145(3):1061–1072

Zhao C-X, Wang Y-q, Wang Y-J, Zhang H-L, Zhao B-Q (2014) Temporal and spatial distribution of PM2.5 and PM10 pollution status and the correlation of particulate matters and meteorological factors during winter and spring in Beijing. Huan Jing Ke Xue 35:418–427

Zheng Y, Lyons T, Ollerenshaw JH, Barnes JD (2000) Ascorbate in the leaf apoplast is a factor mediating ozone resistance in *Plantago major*. Plant Physiol Biochem 38:403–411

Chapter 4
Ozone Biomonitoring, Biomass and Yield Response

Abstract The negative consequences of surface ozone on agricultural crops will be an important threat to global food security in coming years. Several approaches have been adopted from time to time to evaluate the harmful effects of O_3 on crop plants. The technique of O_3 biomonitoring is used to estimate the level of O_3 induced injury in plants. The concept of biomonitoring basically applies two methods, either the estimation of foliar injury symptoms or by analyzing the response of plants using specific chemical protectants which act as antiozonants. Biomonitoring programmes provide us with valuable comparative information regarding the O_3 concentrations at different sites, such that the impact of ambient O_3 is directly measured. The technique is specifically useful in developing countries, where extensive research facilities are not present. Effect of O_3 on plant biomass is an important factor contributing towards the yield of the plants. The biomass accumulation and allocation pattern adopted by the plants in response to O_3 stress not only determines the yield loss but also specifies the cultivar sensitivity/resistivity towards O_3. Different crop loss assessment programmes carried out in different parts of the world have depicted the necessity for further exploring the O_3 induced yield losses in coming times. In addition to programmes like NCLAN and EOTC, several individual experiments have also been conducted to evaluate yield losses in different crop plants. In the past few decades, several models like MOZRAT, MOZRAT 2, have been proposed which have predicted the yield losses of different agricultural crops in coming few decades. As per the modeling results, the regions of south and south-east Asia are specifically prone to greater yield reductions and therefore, there is an urgent need to develop extensive crop loss evaluation programmes in these areas.

Keywords Biomonitoring · Sensitivity · Resistivity · Yield · Crop loss

Contents

© Springer International Publishing AG 2018
S. Tiwari, M. Agrawal, *Tropospheric Ozone and its Impacts on Crop Plants*,
https://doi.org/10.1007/978-3-319-71873-6_4

1   Introduction

Tropospheric Ozone has long been established as an important culprit responsible for causing significant negative effects on crop performance around the globe (Danh et al. 2016; Feng et al. 2015; Pandey et al. 2014; Ainsworth et al. 2012; Avnery et al. 2011a, b). Average O_3 concentration has increased from 20–30 ppb to 30–50 ppb with significant spatial and temporal variations (IPCC 2013). Over the last century, ground level O_3 has increased over four folds (Seinfeld and Pandis 2012) and is still expected to increase at a rate of 0.3 ppb $yr.^{-1}$ (Wilkinson et al. 2011). The continuously increasing concentration of O_3 has seriously affected the physiological and biochemical responses of plants (as discussed in Chap. 3). The damaging effects of O_3 on the biochemical and physiological processes can be manifested in the yield responses, which are depicted as yield reductions. Despite air quality regulations intended to limit O_3 concentration in the troposphere, current ground level O_3 is sufficiently high to suppress the growth and yield of many agricultural and horticultural crops (McGrath et al. 2015; Feng et al. 2015; Wilkinson et al. 2011; Ghude et al. 2014; Emberson et al. 2009; Avnery et al. 2011a, b). Estimated global yield reductions ranged from 2.2% to 5.5% for maize, 3.9–15% for wheat and 8.5–14% for soybean (Avnery et al. 2011a). As per the annual report of IPCC (2013), the most rapid increase in O_3 concentration is currently occurring in south Asia and is also expected to show the maximum increase in near future.

National wise yield loss assessment programmes like NCLAN (National Crop Loss Assessment Programme, USA) and EOTP (European Open Top chamber Programme, Europe) initiated in early 80s and the modeling studies conducted in the last decade were important landmarks for assessing the impact of O_3 on plant yield. The emission models of O_3 precursors in eastern USA, Europe and east Asia suggested that 9–35% of the world's cereal crops are exposed to seasonal O_3 concentrations causing significant reductions in their respective yields (Chameides et al. 1994). Results of 41 studies conducted during NCLAN involving 14 species grown across USA during a 7 year period showed that seven species (cotton, peanuts, spinach, soybean, tomato, turnip and wheat) suffered 10% yield reductions when exposed to an average O_3 concentration (7 hd^{-1}) below 50 ppb (USEPA 1996). It was further shown that more than 20% of Europe crop production land (excluding the grasslands) was estimated to be at risk for yield losses of 5% or more due to O_3 pollution (Mills et al. 2007a).

In USA, modeling studies combining the ground level O_3 concentrations and experimentally derived yield loss function indicated that O_3 levels reduced soybean (*G. max* L. var. Merr) production by 10% in 2005 (Tong et al. 2007). Wang and Mauzerall (2004) also predicted yield reductions of 12–19% for soybean and wheat and 3–5% for rice in China. Climate models were used to forecast that areas like China, Japan, India, central Africa, USA and Indonesia which are responsible for major production of crops like peanuts, rice and soybean will continue to experience much higher O_3 concentrations in near future (IPCC 2013; Emberson et al. 2009; Dentener et al. 2006; Wang and Mauzerall 2004). Feng and Kobayashi (2009) have shown that ambient O_3 concentration with an average of 40 ppb have significantly decreased the yield of major food crops like wheat, rice, soybean, potato etc. by approximately 10% compared to O_3 free air. Feng et al. (2015) evaluated the impact of O_3 on wheat and observed that current and future O_3 levels in China reduced the wheat yield by 6.4–14.9% and 14.8–23%, respectively.

O_3 induced yield losses in crops are preceded by appearance of visible injury symptoms and reductions in growth of the plants (Wan et al. 2013; Pleijel et al. 2014). Visible injury in leaves may not be as biologically significant as yield and growth, but it gives a qualitative assessment of the stress imposed by O_3 (Hayes et al. 2007). The assessment of visible injury has been incorporated in EU/EC International Cooperative Programme on Assessment and Monitoring of Air Pollution effects on Forests (ICP Forests) and on Natural Vegetation and Crops (ICP Vegetation). In North America, visible injury has been included in the local Forest Health Monitoring (biomonitoring) Programme. O_3 induced visible injury symptoms are not only reliable, but also early detectable indicators of O_3 effects on vegetation (Gottardini et al. 2014; Wan et al. 2013; Smith 2012). However, biomonitoring of O_3 through visible injury symptoms face a few limitations as the response of plants may depend upon the plant's genotype (Betzelberger et al. 2010, 2012), phenology (Bagard et al. 2008) and on ecological characteristics of the given site (Bermadinger-Stabentheiner 1996). In addition to the existing O_3 concentration, the above mentioned factors also play significant roles in the development of observable O_3 induced visible injury.

2 Ozone Biomonitoring

2.1 Visible Foliar Injury

Ozone injury symptoms usually occur between the veins on the upper surface of older and middle aged leaves, but may also be bifacial for some species. The severity of injury depends upon several factors including duration and concentration of O_3 exposure, weather conditions and plant genotype. Flecking, stippling, bronzing and reddening on older leaves of the plants are classical symptoms of ambient O_3 injury in plants (Zhao et al. 2011; Sanz and Calatayud 2011; Paoletti et al. 2009). Middleton (1956) and Richards et al. (1958) were the first to report O_3 injury in crop plants. Visible injury in leaves is considered as a valuable tool for assessment of O_3 impacts in field and can be used for identifying areas with high potential risk due to O_3 (Schaub et al. 2010; Matoušková et al. 2010). The O_3 induced foliar injury does not identify the specific level of O_3 present in the ambient air but can give a qualitative assessment of the air quality specifying if the O_3 level is high enough to cause significant damage to plants. The assessment of air quality through the visibly observable responses of plants has led to the development of various biomonitoring programmes that have established that the concentration of ground level O_3

frequently crosses the target value for the protection of human health and vegetation (EU 2002) in several parts of the world. Air quality assessment done via the biomonitoring programmes has played a structural role in framing air quality abatement policies. Gothenberg protocol (UNECE 1999) and NCE Directives (EU 2001) have established certain regulations to check the O_3 concentration by reducing the precursor emissions from traffic and industrial sources.

Certain plant species, also called as bioindicator plants develop O_3 specific injury symptoms which can be used to detect and monitor O_3 stress via biomonitoring programmes. The first reported biomonitoring studies began in northeast and north central USA during 1994–2009 (Smith 2012). This biomonitoring programme provided important regional information on O_3 air quality assessed through measuring O_3 injury (Coulston 2011; Smith et al. 2008). The US Forest Service (USFS) along with US Environmental Protection Agency established this biomonitoring programme which included 24 states divided into two sub regions, the north- east and the north central region and included 450 biosites locations where foliar injury response of O_3 sensitive plants was evaluated for O_3 injury. Out of 46,000 plants, most commonly sampled species were *Rubus allegheniensis*, *Prunus serotina*, *Fraxinus americana*, *Apocymum androsaemifolium*, *Asclepias* sp. etc. O_3 exposure on the biosite location was determined as SUM06 (sum of all hourly average O_3 concentrations >0.06 ppm) and N100 (number of hours of $O_3 > 100$ ppb). O_3 exposure values were grouped under three categories, (i) low (SUM06 < 10 ppm.h; N100 < 5 h), (ii) moderate (SUM06 < 10–30 ppm.h; N100 < 5–30 h), and (iii) high (SUM06 > 30 ppm.h; N100 > 30 h). It was observed that the bioindicator species of sites with high O_3 concentration showed greater injury percent than the species of low O_3 sites. Site level foliar injury index (BI) was calculated for each biosite and it was observed that majority of sites fell into little or no foliar injury group (BI < 5) ranging from 69% in 1994 to 95% in 2003. The percent biosites with most severe injury (BI > 25) reached maximum value in 1994 (10%), 1996 (11%) and 1998 (20%). This biomonitoring programme clearly depicted that foliar injury can be potentially used as an important tool for assessment of O_3 concentration over a large area. It also emphasizes the ability of bioindicator species to detect yearly fluctuations in O_3 concentrations and their usefulness in establishing an exposure-effect relationship. Further, it was also observed that peak O_3 concentration (>100 ppb) during the growing season showed a declining trend, however, this was not the case with seasonal mean O_3 concentration estimated as SUM06 (Smith 2012). The findings of this biomonitoring programme indicated a declining risk of O_3 impacts on north eastern regions as compared to north central United States (Smith 2012).

In Europe, EuroBioNet programme (European Network of Air Quality by use of Bioindicator Plants) was established in 1999 as a network of research institutes and municipal environmental authorities from 12 urban sites in eight EU member states. Along with the other techniques, this programme employed extremely O_3 sensitive tobacco (*Nicotiana tabaccum* L) cultivar Bel W3 to access occurrence of phytotoxic effect of O_3 at different selected sites. As a part of EuroBioNet programme, Klumpp et al. (2006) evaluated the response of O_3 sensitive tobacco cultivars at 100 biomonitoring sites in urban centres for 3 years and tried to establish a relationship between O_3 concentration and O_3 induced visible foliar injury on tobacco. Results of this experiment showed highly significant correlation between foliar injury level and various O_3 exposure indices used in this study (AOT20 and AOT40) (Klumpp et al. 2006). Ribas and Peñuelas (2003) also recorded significant correlation between AOT20 and AOT40 values and O_3 induced foliar injury on tobacco cultivar Bel W3 in a biomonitoring study at a rural site in Spain.

An analysis of different biomonitoring studies has shown variations in the observed relationship between ambient O_3 concentrations and symptomatology of bioindicator plants. As compared to the large area networks, small scaled studies showed a near linear relationship between O_3 concentration and its effect on visible foliar injury (Cuny et al. 2004; Kostka- Rich 2002; Ribas and Peñuelas 2002). This observed variation can be explained by the homogenous meteorological conditions in small study area as compared to large area (Klumpp et al. 2006). In addition to this, there exists a possibility that the actual O_3 values affecting the plant may differ from the measured concentration. It has now been well established that O_3 flux through the stomata and cumulative O_3 uptake rather than ambient O_3 concentration determine the O_3 induced effects on plants (Mills et al. 2011, 2016). Results of the EuroBioNet programme demonstrated that O_3 pollution and O_3 induced effects increased along a gradient from northern Europe to central and southern Europe, however, the topographic features and distribution of O_3 precursor emissions greatly influenced the O_3 concentrations and their effects on local scale (Schaub and Calatayud 2013; Klumpp et al. 2006).

In addition to the extensive biomonitoring programmes, several other studies have also analyzed the O_3 induced visible foliar injury in plants in different parts of the World (Kohut 2005; Paoletti et al. 2009; Sanz and Calatayud 2011; Wan et al. 2014; Feng et al. 2014). Using biomonitoring techniques, visible injury was recorded in O_3 sensitive tobacco cultivar Bel W3 grown at urban, suburban, semi rural and rural locations in and around the city of Lahore (Kafiatullah et al. 2012). Monitoring studies from 1993–94 to 2006 showed that the highest O_3 levels were found at the rural (seasonal mean ranging from 73.9–99.7 ppb) and minimum at urban centre (54–67.7 ppb) (Kafiatullah et al. 2012). The visible injury in tobacco cultivar Bel W3 showed a linear relationship with O_3 concentration, where the highest leaf injury index of 18–27% was recorded at the rural site and lowest of 5–7% at the urban centre (Kafiatullah et al. 2012). These findings were consistent with those of Agrawal et al. (2003) which also recorded the strongest O_3 induced visible foliar injury symptoms at a suburban site having higher O_3 concentration. Transect studies done in and around Varanasi city showed that the magnitude of O_3 induced negative effects on plant's performance depicted a direct positive correlation with O_3 concentration in mung bean (*Vigna radiate*), palak (*Beta vulgaris*), wheat (*Triticum aestivum*) and mustard (*Brassica oleracea*) (Agrawal et al. 2003). Calatayud et al. (2007) also recorded foliar injury symptoms in tobacco cultivar Bel W3 grown at 7 sites of Eastern Spain, which included 4 traffic exposed urban sites, a large urban garden and a suburban and a rural centre. The observed foliar injury showed a moderate

Fig. 4.1 Foliar injury symptoms in different cultivars of clover (*Trifolium alexandrium* L.) under elevated ozone (Source: Chaudhary and Agrawal 2013)

correlation with O_3 concentration. Urban centres with intense traffic showed less O_3 injury as compared to less traffic exposed sites (Calatayud et al. 2007).

Appearance of visible foliar injury symptoms depends not only upon O_3 concentration but also upon the plant genotype. This view is supported by the findings of Chaudhary and Agrawal (2013) who observed that the magnitude of visible injury showed variations in different cultivars of clover (*Trifolium alexandrium* L.) depending upon their sensitivity/resistivity towards O_3. Six cultivars of clover were exposed to mean elevated O_3 concentration of 57.5 ppb and it was observed that the cultivar Wardan, which was most sensitive to O_3 showed maximum percentage of foliar injury in form of small pale yellow and brown flecks on the upper surface that turned into bifacial necrosis upon extended exposure to O_3 (Fig. 4.1) (Chaudhary and Agrawal 2013). After Wardan, cultivars Bundel and JHB-146 also followed the same pattern of development of foliar injury upon O_3 exposure, but injury percentage was lesser than Wardan. In cultivars Fahili, Saidi and Mescavi, which are O_3 resistant cultivars, injury symptoms appeared during later growth stages and the percentage of foliar injury was less than the other cultivars (Chaudhary and Agrawal 2013). Similar kind of variation in response to O_3

induced foliar injury was recorded in two genotypes of white clover (*Trifolium repens* L. cv Regal) NC-S (O_3 sensitive) and NC-R (O_3 resistant) upon O_3 exposure to mean O_3 concentration varying between 28–61 ppb in Bangladesh (Islam et al. 2007). In this experiment, O_3 sensitive genotype NC-S showed foliar injury symptoms, whereas no symptoms were recorded in O_3 resistant genotype NC-R (Islam et al. 2007). These two biotypes of *Trifolium repens* L. were also used for biomonitoring experiments in 17 European countries from 1996–2006 in an international programme coordinated by ICP vegetation (Mills et al. 2011). In addition to this, Mills et al. (2011) compiled the data (644 studies) for O_3 induced visible injury in crops (27 species), shrubs (49 species) and semi- natural vegetation (95) species in Europe and observed that O_3 concentrations were sufficiently high to cause damage to vegetation. The study showed evidence of O_3 injury even at the sites where AOT40 values were found to be low. However, the degree of O_3 injury at these sites as determined by O_3 flux which could be moderate to high depending upon the climatic conditions being favourable or unfavourable for stomatal uptake. This study by Mills et al. (2011) provided important rationale data to support the use of flux based metrics rather than exposure based ones.

Wan et al. (2014) studied the O_3 injury symptoms on native and cultivated trees, shrubs and herbs at ten different urban and rural sites in and around Beijing in China. It was observed that 19 species of native and cultivated trees, shrubs and herbs at the selected sites exhibited symptoms of O_3 injury in form of upper surface stipple (Figs. 4.2, 4.3 and 4.4) (Wan et al. 2014). The study further reported significant O_3 injury in *Ailanthus altissima* and *Hibiscus syriacus*, which are already established as important bioindicator plants for ambient O_3 (Seiler 2012; Sanz and Calatayud 2011). In addition to these, O_3 injury was reported for the first time in nine species, namely *Cassia tora*, *Phorbitis purpurea*, *Canavalis gladiate*, *Amgdalus triloba*, *Kerria japonica* var. plenifolia, *Hyptis suaveolens*, *Vigna unguiculata*, *Sophora aureus*, and *Juglans regina* (Wan et al. 2014). This work reported visible injury in *Populus tomentosa* for the first time. Other species of the genus *Populus* are already known to be O_3 sensitive (Kohut 2005). Figures 4.2, 4.3 and 4.4 show the visible injury symptoms reported for selected sensitive trees, shrubs, herbs and crop plants (Wan et al. 2014). Frequency of foliar injury symptoms were 56.3% (18/32) for trees, 15.6% (5/32) for shrubs, 21.9% (7/32) for herbs and 6.3% (2/32) for crop plants (Wan et al. 2014). Mean O_3 concentration during the experimental period was recorded through passive sampler and varied from 22.5 to 48.1 ppb with highest O_3 concentration recorded at locations outside the city centre. The O_3 injury symptoms, however, did not correlate with the recorded O_3 concentrations at different sites. It was reported that the site with high O_3 concentration showed lesser O_3 injury which can be attributed to the existing environmental conditions that were unfavourable for O_3 uptake by plants (Wan et al. 2014).

Feng et al. (2014) also confirmed that O_3 concentrations in and around Beijing were sufficiently high to induce visible foliar injury in plants. O_3 induced foliar injury symptoms were recorded in 28 species and were more prominent in rural and mountain areas in northern Beijing, downwind from the city and less frequent in

Fig. 4.2 The injury symptoms on sensitive trees: (**a**) *Ailanthus altisima* (**b**) *Populus tomentosa* (**c**) *Rhus typhina* (**d**) *Ulmus pumila* (Source: Wan et al. 2014)

city gardens. Based upon the results, it was found that in addition to *Phaseolus vulgaris* which is a well recognized bioindicator for O$_3$, other types of beans belonging to the genera *Canavalia*, *Vigna* and other species of *Phaseolus* also showed distinctive O$_3$ injury symptoms (Fig. 4.5). Agricultural crops like *Citrullus lanatus*, *Vitis vinifera*, *Abelmoschus esculentus*, *Arachis hypogea*, *Cucurbita pepo*, and *Lagenaria siceraria*, trees like *Ailanthus altissima*, *Fraxinus chinensis*, *Koelreuteria paniculata*, *Ampelopsis humufolia* and *Sambuscus williamsii* and ornamental plants like rose of Sharon, Japanese morning glory and black locust also showed distinct injury symptoms in the presence of O$_3$ (Fig. 4.5) (Feng et al. 2014). As a part of biomonitoring programme of ICP Vegetation, Sanz and Calatayud (2013) studied the foliar injury symptoms of European Forest species and reported

Fig. 4.3 The injury symptoms on sensitive shrubs: (**a**) *Amygdalus triloba* (**b**) *Salix leucopithecia* (**c**) *Hibiscus syriacus* (**d**) *Kerria japonica* var. pleniflora (Source: Wan et al. 2014)

prominent visible injury symptoms on the leaves of *Ailanthus altissima, Citrullus lantanus, Fraxinus chinensis, Fraxinus rhynchophylla, Phaseolus vulgaris, Sambucus williamsii,* etc. (www.ozoneinjury.org).

ICP Forests Intensive Monitoring Network launched a specific pan-European programme for assessment, validation and mapping of O_3 injury on vegetation. Through this programme, over a period of 2002–2014 across Europe, data were recorded and analyzed from 285 species on 169 plots across 19 countries. 55% plots were symptomatic and 26% of the 285 assessed species developed visible injury. Out of the symptomatic species *Fagus sylvestica* showed maximum percentage of injury (40.1%), while percentage of foliar injury was as low as 16.9% in *Fraxinus excelsior* and 12.9% in *Rubus fruticosus* (Michel and Seidling 2016). The data

Fig. 4.4 The injury symptoms on sensitive crops and herbs: (**a**) *Cassia tora* (**b**) *Ipomoea nil* (**c**) *Canavalia gladiate* (**d**) *Helianthus sannuus* (**e**) *Pharbitis purpurea* (**f**) *Hyptis suaveolens* (Source: Wan et al. 2014)

recorded during this programme clearly indicates that O_3 still occurs at the levels harmful to forest vegetation. Visible leaf injury symptoms have also been observed in over 60 species of crops, wild flowers, shrubs and trees over a period of 2007–2015 in over 19 countries from Europe and North America (Harmens et al. 2016). Under the ICP vegetation biomonitoring programme, members of ICP Vegetation

Fig. 4.5 Ozone symptoms in native plants: (1) *Ailanthus altissima*, (2) *Ampelopsis humulifolia*, (3) *Fraxinus rhynchophylla*, (4) *Pinus bungeana*; Ornamental plants: (5) *Robinia pseudoacacia* (6) *Hibiscus syriacus*; Crops: (7) *Canavalia gladiata* (8) *Vigna unguiculata* var heterophylla, (9) *Benincasa pruriens*, (10) *Luffa cylindrica*, (11) *Citrullus lanatus*, and (12) *Vitis vinifera* (Source: Feng et al. 2014)

have been recording the presence of visible O_3 injury symptoms on plants in field and natural habitats (http://icpvegetation.ceh.ac.uk/record/index).

2.2   Use of Ethylenediurea

In addition to the visible foliar injury, O_3 biomonitoring programme also utilized chemical protectants which help in the evaluation of O_3 induced injury in plants. Ethylenediurea ($C_4H_{10}N_4O_2$), abbreviated as EDU is an antiozonant which is structurally N-[−2-(2- oxo-1- imidazolidinyl)ethyl]-N´- phenylurea (Wat 1978), has emerged as a versatile research tool for evaluating O_3 injury in plants. Its specificity towards O_3 has made it an important biomonitoring tool for assessing O_3 injury

(Manning et al. 2011; Singh et al. 2015; Agathokleous et al. 2016a, b, c, d). Carnahan et al. (1978) reported for the first time that EDU was effective in providing partial protection to the plants against O_3 injury. In the last decade, extensive researches have been done to access the degree of protection rendered by EDU to O_3 stressed plants (Hoshika et al. 2013; Paoletti et al. 2014; Carriero et al. 2015; Agathokleous et al. 2016a, b, c, d). EDU is used as a biomonitoring tool at doses usually ranging from 200–400 mg L^{-1} (Feng et al. 2010). The mode of action of EDU is still debated, however, in the recent studies, at higher doses (>800 mgL^{-1}), an increase in foliar N content of the EDU treated plant (*Salix sachalinensis* L. Schm) was observed (Agathokleous et al. 2016c). Increase in foliar nitrogen content in EDU treated plants led to the speculation that N may play an important role in EDU mediated partial amelioration of O_3 stressed plants. However, Godzik and Manning (1998) applied 300 mg L^{-1} EDU and corresponding amount of N as urea (70 mg L^{-1}) and phenylurea (159 mg L^{-1}) to the leaves of tobacco plants (*Nicotiana tabacum* L. Bel W3) and observed that N was not responsible for the mechanism of EDU protection in plants. In O_3 sensitive bean (*Phaseolus vulgaris* L var. S-156) treated with 300 mg L^{-1} EDU, a slight increase in N content was recorded only a day after EDU treatment, after which the N content declined (Manning et al. 2011). Paoletti et al. (2007a, b; 2008) observed that six EDU treatments (450 mg L^{-1}) as gravitational trunk infusions, did not contribute to the leaf N content of *Fraxinus excelsior* L. On the basis of these studies, the role of N in EDU is ambiguous and requires more experimentation before EDU bound N can be established to act as fertilizer (Agathokleous et al. 2016c).

EDU does not enter the cell but remains confined to the foliar apoplastic spaces (Pasqualini et al. 2016) and hence is systemic in nature (Paoletti et al. 2009; Weidensaul 1980). Its retention time in the leaf varies from 8 days (Paoletti et al. 2009) to 21 days, depending upon the soil fertility conditions (Agathokleous et al. 2016d). As such, repeated application of EDU (after every 9 days) is recommended for the treatment to be effective. Depending upon the O_3 concentration and the sensitivity of the test plant towards O_3, EDU treatment has given some promising results in form of reduced visible injury (Singh et al. 2015), increased plant growth and biomass (Rai et al. 2015; Yuan et al. 2015; Xin et al. 2016) and significant yield increments of O_3 stressed plants (Feng et al. 2015; Xin et al. 2016; Tiwari and Agrawal 2010; Tiwari et al. 2005), which make it suitable as an important biomonitoring tool for assessing O_3 injury. From the time of Carnahan's first EDU report in 1978 till 2011 (Manning et al. 2011), 115 papers have been published, analyzing the role of EDU in ameliorating O_3 injury. The results of these studies have found a positive effect of EDU on plants (Manning et al. 2011). Meta analytical study done by Feng et al. (2010) also indicated the usefulness of EDU in partial amelioration of O_3 stress in plants. EDU, as such can be used as a biomonitoring tool on the basis of the following features:

  (i)   Evaluating the role of O_3 in occurrence of foliar symptoms in plants.
 (ii)   Determining the effect of ambient O_3 on growth and yield of plant.
(iii)   Screening plant sensitivity to O_3 under ambient conditions.

The biomonitoring aspect of EDU can also be utilized in evaluating the effect of global warming on plant productivity. Warmer climate indicates increases in CO_2 and O_3 concentration in the atmosphere. Using EDU in the field can be helpful for the assessment of CO_2 effects in the absence of O_3. This experimental approach will help us to determine if O_3 will cancel out the beneficial effect of CO_2 on plants in the coming time. EDU, as a biomonitoring tool gains more significance in areas where O_3 monitoring is not feasible due to certain technical problems like disruptive electricity supply, unavailability of monitoring instruments, etc. (Paoletti et al. 2009). O_3 injuries at remote sites of India (Agrawal et al. 2005; Tiwari et al. 2005) and Pakistan (Wahid et al. 2001) were determined using EDU. Studies analyzing the sensitivity of plants using EDU suggested that sensitive genotypes responded more positively to EDU treatment than the O_3 tolerant ones (Tiwari et al. 2005; Pandey et al. 2014).

3 Effects of Ozone on Plant Biomass

O_3 clearly affects the growth of the plants which is evident through reduced biomass of the O_3 exposed plants. Assessment of O_3 induced biomass reduction is significant as in many cases biomass variations are manifested in form of yield reductions in O_3 exposed plants (Sitch et al. 2007; Collins et al. 2010). O_3 effect on the economic yield of a plant largely depends upon the partitioning of biomass between reproductive and non reproductive parts of the plant (Pleijel et al. 1995; Leisner and Ainsworth 2012). The relationship between grain yield, biomass partitioning and biomass accumulation can be represented by (Pleijel et al. 2014):

$$Y_G = H_I B_A$$

where, Y_G is the grain yield, H_I is the harvest index and B_A is the above ground biomass at harvest.

The significant effect of O_3 on H_I can be estimated by post anthesis high sensitivity to O_3 in annual crops such as wheat (Pleijel et al. 1998). O_3 is known to speed up the process of senescence in plants post anthesis stage which shortens the grain filling period (Gelang et al. 2000). This results in smaller grain biomass in relation to total biomass (H_I) representing the reduced efficiency of conversion of accumulated biomass to yield. O_3 exposure thus disturbs the biomass yield interconversion relationship in plants (Pleijel et al. 2014). Using the data of 22 O_3 exposure experiments with field grown wheat in Europe, European Scale Chemical Transport Model (EMEP.MSC- West) estimated a loss of 9% in above ground biomass and 14% in yield (Pleijel et al. 2014). Based upon this data, it was suggested that O_3 induced loss of agroecosystemcarbon storage should be estimated with respect to reduction in biomass rather than the yield reduction which overestimated the effect of O_3 on carbon sequestration (Pleijel et al. 2014).

Gerosa et al. (2015) studied the response of Holm Oak (*Quercus ilex*) exposed to four different levels of O_3 in OTCs during one growing season. The O_3 treatments used in the experiment were charcoal filtered (CF), non filtered ambient O_3 concentration (NF), NF + ambient O_3 concentration increased by 30% (NF+) and NF + ambient O_3 concentration increased by 74% (NF++). This study evaluated the biomass losses on roots, stem and leaves with respect to O_3 exposure (AOT40) and phytotoxical O_3 dose (POD_1) absorbed. The concentrations of O_3 (AOT40/POD_1) were 13,076 ppb.h/17.2 mmol m^{-2}, 1474 ppb.h/34.2 mmol m^{-2}, 26,554 ppb.h/40.7 mmol m^{-2} and 55,611 ppb.h/53.4 mmol m^{-2}, respectively, for NF, CF, NF+ and NF++ O_3 treatments. This study recorded reductions of 4 and 5.25% of the total biomass for each increased step of 10,000 ppb.h of AOT40 and 10 mmol m^{-2} of POD_1, respectively (Gerosa et al. 2015). The overall reduction of dry biomass was greater for roots than for stem and leaves, with 27% significant decrease in NF++ treatment as compared to control (CF treatment), whereas the stem dry biomass showed a significant reduction of 16% (Gerosa et al. 2015). This study showed that both dose effect and exposure effect relationship between O_3 concentration and biomass reduction had similar regression values indicating similar response of plant biomass with respect to both the metrics (Gerosa et al. 2015). This result, however, is in contradiction with the earlier findings which suggest that O_3 flux metrics better explained the biomass reduction than O_3 exposure metrics. On the basis of these results, Gerosa et al. (2015) suggested a critical level of 70 ppb for Holm Oak protection which resulted in a total biomass reduction of 4%. This POD_1 value is about twice the critical level currently set by UNECE for protection of broad leaved deciduous species (4 mmol m^{-2} for beech and birch at 4% of total biomass reduction), but well below the critical level suggested for evergreen coniferous forests (8 mmol m^{-2} for Norway Spruce when the biomass decrease is 2%). On the basis of observed biomass reductions relative O_3 sensitivity of *Q. ilex* was found to be set between that of deciduous broad leaved and evergreen coniferous species (Gerosa et al. 2015).

Another experiment was conducted in central Spain to evaluate the effect of O_3 on biomass of 6 selected species in an annual pasture community (Calvete-sago et al. 2014). In an OTC experiment, 3 legumes (*Trifolium striatum*, *Trifolium cherleri* and *Ornithopus compressus*), 2 grasses (*Briza maxima* and *Cynosurus echinatus*) and 1 forb (*Silene gallica*) were exposed for 40 days to four O_3 treatments viz. filtered air (CF), non filtered air with ambient O_3 (NF) and NF air supplemented with 20 and 40 ppb O_3. The negative effect of O_3 on pasture community was evident by a reduction of 25% in green biomass and up to 21% in above ground biomass (Calvete-sago et al. 2014). The result of this experiment suggest that the annual pasture community seemed to be more sensitive to O_3, where the O_3 induced significant effects were observed after a few weeks of O_3 exposure (Calvete- ago et al. 2014). Contrary to this, Volk et al. (2011) did not find any significant response of perennial alpine grassland even after 5 years of O_3 exposure.

Biomass allocation to different plant parts during reproductive phase plays an important role in determining O_3 induced yield reductions, which largely depends on efficacy of compensatory reproductive processes (Black et al. 2010). It has been observed that a moderate increase in O_3 concentration is sufficient to cause a decrease in plant's reproductive growth characteristics, the effect showing prominent increase with increasing O_3 concentration (Black et al. 2000). Leisner and Ainsworth (2012) evaluated the effect of O_3 on reproductive development of plants by analyzing data from 128 peer reviewed studies from 1968 to 2000 using meta analytical techniques and observed that ambient O_3 significantly decreased the fruit number and weight by 9 and 22%, respectively, as compared to charcoal filtered air. It was reported that O_3 did not significantly alter the inflorescence number, flower weight and number of flowers in several plant species (Leisner and Ainsworth 2012). Thwe et al. (2014), however, observed significant reductions in flower and fruit numbers in tomato (*Solanum lycopersicum* Mill ver "Look Tor") exposed to O_3 concentrations of 178.5 and 255 ppb. Reduction of 28.3% in number of fruits plant^{-1} and a corresponding 29% reduction in yield (fresh weight of fruits plant^{-1}) were recorded in pepper (*Capsicum annuum* L.) grown at an O_3 concentration of 78 ppb (Taia et al. 2013). Studies have shown that the negative effect of O_3 on reproductive characteristics can be attributed to the effect on carbon assimilation and allocation within the plants (Leisner and Ainsworth 2012). Thwe et al. (2014) observed a direct correlation between altered responses of reproductive parameters (number/weight of flowers/fruits plant^{-1}) and variation in carbon acquisition and allocation within the plants exposed to O_3. Grantz and Yang (2000) showed that O_3 induces an allometric shift in carbohydrate allocation of Pima cotton (*Gossypium barbadense* L.) due to direct effect of O_3 stress on phloem loading.

4   Ozone Induced Yield Response

Significant yield reductions of agricultural crops is a well cited consequence of rising tropospheric O_3 (Avnery et al. 2011a, b; Ainsworth et al. 2012; Cotrozzi et al. 2016; Yi et al. 2016). Due to transport of O_3 precursors across National boundaries and continents, rising emissions of O_3 in the Asian continent play an important role in increasing the hemispheric background O_3 concentration (Cooper et al. 2014). Estimating the loss of agricultural production from ground level O_3 provides us with valuable information for understanding the potential benefits of reducing O_3 and projecting future food supply (Burney and Ramanathan 2014). Predicted trends suggest that ground level O_3 in south Asia has shown maximum increments and will continue to follow the same trends until 2050 (IPCC 2013). Therefore a thorough understanding of crop and cultivar response to O_3 and the incorporation of these responses into crop production models are important to evaluate the potential impact of O_3 on food supply in different parts of the world. Numerous controlled experiments using air filtration techniques were carried out in different parts of the world to evaluate the O_3 induced yield losses. Based on the large scale experimentation studies of NCLAN conducted in United States in 1980s (Heagle 1989; Heck 1989), the US Environmental Protection Agency (EPA) predicted that yield of about 1/3 of US crops was reduced by 10% due to ambient O_3 concentration (EPA 1996). Results from European Open Top Chamber Programme (EOTP) in 1990s, similarly observed that Europe may be losing more than 5% of their wheat yield due to O_3 exposure (Mauzerall and Wang 2001). Open Top Chamber (OTC) field studies conducted in Europe and USA during these programmes established crop specific concentration response (CR) functions that predict the relation between the yield reduction of a crop and different levels of O_3 exposure.

Although the developing countries of south and east Asia lack extensive yield loss assessment programmes like NCLAN and EOTC, potential risks to ambient O_3 to agricultural production have been evaluated through small scale field experiments and modeling studies (Tiwari et al. 2006; Emberson et al. 2009; Avnery et al. 2011a, b; Debaje 2014; Feng et al. 2015; Danh et al. 2016; Yonekura and Izuta 2017). Based upon the results of modeling studies conducted in Asia, mean monthly O_3 concentration frequently reaching 50 ppb has been recorded during the growing season of important agricultural crops (EANET 2006). Global photochemical models further project significant rise in O_3 concentration by 2030 (Dentener et al. 2006). The potential risk to agricultural crops as posed by O_3 may become more relevant for Asian regions which are further expected to register large increase in population by 2030 (Tiwari 2017; Sarkar et al. 2017).

Several workers have used the dose response relationship derived from the NCLAN and EOTP programmes to quantify the yield losses for economic crop loss estimates in Asian scenario (Aunan et al. 2000; Wang and Mauzerall 2004; Emberson et al. 2009). Wang and Mauzerall (2004) estimated yield loss up to 13 and 23%, which was expected to increase to 16 and 35% for cereals and soybean, respectively, during 2020 and translated the yield loss estimates into economic loss of approximately US$ 5 billion for wheat, rice, maize and soybean growing in China, Japan and South Korea. However, the studies relying upon the dose response relationship based on North American and European experimental data do not provide a clear picture to the yield response of plants in Asian context. Fuhrer and Booker (2003) suggested that factors such as climate, crop phenology, agricultural management practices and pollutant exposure pattern altered the plant response to O_3. In order to check the hypothesis that the dose response relationship based on North American data provide an accurate estimation of crop responses to O_3 in Asia, Emberson et al. (2009) pooled the data of a number of experimental studies done on a wide range of crop species and cultivars in several Asian countries like India, China, Japan and Pakistan. It was convenient to use the data of North American studies for comparison with the Asian studies since their method of characterizing O_3 exposure is nearly similar in both the cases i.e. as M7 (7 h) or M12 (12 h). In contrast, the AOT40 concept, developed in Europe makes the use of hourly O_3 concentration data, which was unavailable from the Asian studies (Emberson et al. 2009). The data compiled by Emberson et al. (2009) was based upon studies on wheat (*Triticum aestivum* L) (10 studies), rice (*Oryza sativa* L) (7 studies), soybean (*Glycine max* L) (3 studies) and mungbean (*Vigna radiate* (L) R Wilczek) (3 studies) which included three experimental approaches viz. fumigation, filtration and chemical protectant (EDU) studies. These studies used 13 wheat (8 studies with winter wheat and 5 with spring wheat), 10 rice, 4 soybean and 3 mungbean cultivars.

A comparative analysis of the results indicates that the O_3 effects on different crops were strongly influenced by the experimental method. In case of rice and wheat, filtration experiments showed a higher sensitivity of both the crops with relative yield losses up to 50% in both wheat (between 36 and 72 ppb M4- M8 ambient O_3) and rice (between 33 and 60 ppb M4-M8 ambient O_3). In soybean, the data indicate that fumigation studies have far reduced sensitivities with yield losses in the range 10–15% compared to those of the EDU studies where yield losses varied between 30–65% (Emberson et al. 2009).

Emberson et al. (2009) compared the response of Asian wheat varieties used in the study with the North American dose response relationship as obtained through NCLAN programme and observed that the Asian spring wheat varieties were more sensitive than North American spring wheat varieties. However, the Asian winter wheat varieties showed variable response; a few studies (4 data points) depicted greater resistance while the remaining 12 data points (from 7 studies) showed greater sensitivity as compared to North American NCLAN dose response relationship based on 4 varieties (Heck 1989). In case of rice, all cultivars (except one) used in the study showed higher sensitivity to O_3 as compared to North American dose response relationship (Adams et al. 1989). However, interpretation of legume sensitivity when compared with North American dose response relationship (Lesser et al. 1990) does not follow a specific trend as the fumigation experiments showed a lower sensitivity and filtration and EDU experiments showed higher sensitivity compared to the North American dose response relationship (Emberson et al. 2009). On the basis of the Asian data for crop yield loss, the ranking for increasing crop sensitivity to O_3 is in the order of rice < wheat < legume (Emberson et al. 2009), which is consistent with the other studies (Heck et al. 1988; Mills et al. 2007b). Through this comparison between North American and Asian O_3 concentration and response relationship, it was clear that this function underestimated the effect of O_3 on crop yield in Asia.

Van Dingenen (2009) modeled O_3 concentration through Chemical Transport Model (TM5) and calculated Relative Yield Loss (RYL) for the year 2030 based on "current legislation scenario (CLE)" assuming that currently approved air quality legislations will be fully implemented by 2030. Depending upon the O_3 exposure metrics used (AOT40 or M7), global RYL for the year 2000 was estimated to range between 7–12% for wheat, 6–16% for soybean, 3–4% for rice and between 3–5% for maize (Van Dingenen 2009). Under the 2030 CLE scenario, the global increase in RYL for wheat (4%), soybean (0.5%), maize (0.2%) and rice (1.7%) is expected, with the countries of South Asia being most severely affected (Van Dingenen 2009). This study suggests that wheat and rice are most severely affected crops with 40% losses occurring in China and India (Van Dingenen 2009). Aunan et al. (2000) combined O_3 dose response relationship from European and American experiments in an global three dimensional photochemical tracer/transport model (CTM) and predicted that by 2020, O_3 induced yield reductions in China would be 18–21% for soybean, 29.3% for spring wheat, 13.4% for winter wheat, 7.2% for corn and 5% for rice. Wang and Mauzerall (2004) used MOZRAT-2 (Model of Ozone and Related Chemical Tracers, Version 2) and calculated that O_3 concentration caused yield loss of 8% in 1990 (based on M7/M12 indices) and predicted that four grain crops (wheat, rice, soybean and corn) would suffer production losses estimated to be 47.4 million metric tons in China.

Based upon the estimations with high resolution chemical transport model coupled with Regional Emission Inventory in Asia (REAS), Tang et al. (2013) estimated the RYL for wheat in India and China for the years 2000 and 2020 using dose response function based on exposure metrics AOT40 and metric associated with O_3 flux, PODy. RYL for wheat in 2000 was estimated to be in the range of 6.4–14.9% and 8.2–22.3% for China and India, respectively (Tang et al. 2013). Based upon both O_3 metrics, an increase in RYL from 2000–2020 in the range of 8.1–9.4% and 5.4–7.7% for China and India, respectively is predicted (Tang et al. 2013). Debaje (2014) estimated the yield losses of two major crops of India (winter wheat and rabi rice) for the period between 2002–2007, on the basis of two indices of O_3 exposure (AOT40 and M7). Winter wheat showed yield loss of 5–11% (using M7 index) and 6–30% (using APT40 index). In case of rice, the reductions were 3–6% and 9–16% through M7 and AOT40 O_3 indices, respectively.

Avnery et al. (2011a) estimated the global yield reductions of three staple crops for the year 2000 due to surface O_3 exposure using hourly O_3 concentrations stimulated by MOZRAT-2 and observed yield reductions ranging from 8.5–14% for soybean, 3.9–15% for wheat and 2.2–5.5% for maize, depending upon the metric of O_3 exposure (M12 or AOT40). Avnery et al. (2011a) calculated the Relative Yield Lost (RYL), Crop Production Loss (CPL) and Economic Loss (EL) for soybean, maize and wheat for different regions of the world based upon O_3 concentration simulated by MOZRAT -2. Using the two O_3 exposure metrics, O_3 induced RYL of wheat is highest in Bangladesh (15–49%), Iraq (9–30%), India (9–30%), Jordan (9–27%) and Syria (8–25%) (Fig. 4.6). In case of soybean, RYL is estimated to be greatest in Canada (27–28%) followed by Italy (24–27%), South Korea (21–25%), China (21–25%) and Turkey (16–23%) (Fig. 4.6). For maize, the RYL was comparatively smaller, with highest losses occurring in Democratic Republic of Congo (DRC) (7–13%), Italy (7–12%), Canada (6–11%), South Korea (4–9%) and Turkey (4–9%). Table 4.1 shows the regional and global estimates of RYL based upon M7, M12 and AOT40 metrics (Avnery et al. 2011a). Wheat yield reductions were found to be most significant in south Asia (17.4%), whereas for soybean, reductions were maximum in EU-25 (25.6%) followed by east Asia (22.8%). RYL for maize was more evenly distributed with highest reduction recorded in east Asia (5.9%) followed by south Asia and EU-25 (5.7%) each (Avnery et al. 2011a). Worldwide CPL was estimated to be between 21–93 million metric tons (Mt) for wheat, 13–32 Mt. for maize and 15–26 Mt. for soybean (Avnery et al. 2011a). CPL of wheat showed large variations due to the fact that this crop appears resistant according to the M12 metric, but extremely sensitive to O_3 according to AOT40 index. Global CPL for all the three crops was found to be in the range of 79–121 Mt. (Avnery et al. 2011a). Global Economic Loss (EL) due to yield reductions of all the three crops ranged from $ 11–18 bilion, with soybean accounting for $ 2.9–4.9 billion. The greatest EL was found in US ($3.1 billion) followed by China ($ 3 billion) and India ($2.5 billion) (Avnery et al. 2011a).

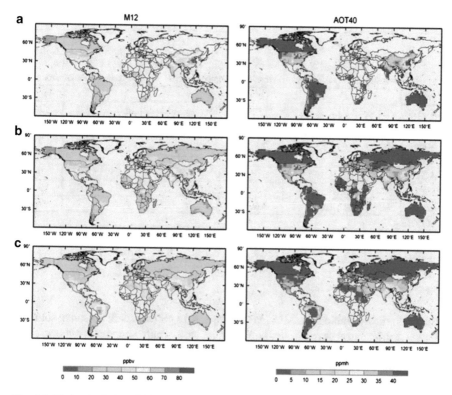

Fig. 4.6 National relative yield loss according to M12 and AOT40 metrices for (**a**) soybean (**b**) maize (**c**) wheat (Source: Avnery et al. 2011a)

Unfortunately, the MOZRAT-2 used by Avnery et al. (2011a) overestimated the O_3 exposure in US especially in the north and south eastern parts of the country by upto 22 ppbv, during the summer growing season. Similarly the over prediction of O_3 by MOZRAT-2 in northern India may lead to overestimation of agricultural losses in this region, especially for wheat. However, since the Asian wheat cultivars were predicted to be more sensitive to O_3 than the cultivars from US (Emberson et al. 2009), the error caused by the model over prediction may be counteracted. In southern India, however, MOZRAT-2 performs well, so the western O_3 concentration – response relationship may lead to an underestimation of crop loss. In China, however, O_3 is slightly underestimated by MOZRAT-2 and the regional crop cultivars also exhibit greater sensitivity to O_3 exposure. MOZRAT -2 missed some of the seasonal trends in Japan as well, under predicting O_3 concentrations in April by ~20 ppbv and over predicted in fall by up to ~15 ppbv. In southern Africa also, the agricultural losses were highly biased due to overestimation of surface O_3 (Avnery et al. 2011a). The data from Australia and New Zealand also faced a similar problem of O_3 over prediction by MOZRAT-2 during the dry season, however, the model performed well for the remaining part of the year. The model also underestimated O_3 in central Europe by ~ 5–17 ppbv during the first half of the year.

Table 4.1 Estimated Relative Yield Loss (RYL) (%) for the year 2000 due to ozone exposure

	World	EU-25	USSR & E. Europe	North America	Latin America	Africa & Middle East	E. Asia	S. Asia	ASEAN & Australia
WHEAT									
AOT40	15.4	12.1	11.4	11.0	5.9	20.1	16.3	26.7	1.0
M7	3.9	3.3	2.4	2.6	1.5	5.9	3.3	8.2	0
Mean	**9.6**	**7.7**	**6.9**	**6.8**	**3.7**	**13.0**	**9.8**	**17.4**	**0.5**
MAIZE									
AOT40	2.2	3.5	2.3	2.0	0	0.6	3.8	3.4	0.3
M12	5.5	7.9	6.5	5.1	2.1	2.5	8.0	8.0	2.4
Mean	**3.9**	**5.7**	**4.4**	**3.6**	**1.2**	**1.6**	**5.9**	**5.7**	**1.4**
SOYBEAN									
AOT40	8.5	23.9	–	12.0	0.2	2.0	20.9	3.1	0
M12	13.9	27.4	–	16.9	6.3	9.8	24.7	13.2	3.7
Mean	**11.2**	**25.6**	–	**14.4**	**3.3**	**5.9**	**22.8**	**8.2**	**1.9**

Source: Avenry et al. (2011a)

Based on the available data, MOZART-2 appears to perform well in China, southern India, north/central Africa, southern Africa, and south America (Avnery et al. 2011a). Although the reasons for this bias remain somewhat unclear, possible explanations include the coarse resolution of global CTMs, as well as potential issues related to environmental chemistry of the area, isoprene emissions and oxidation pathways, and the discharge of elevated emission point sources into the model surface layer (Horowitz et al. 2007; Reidmiller et al. 2009).

The variable sensitivities of different crop cultivars and different experimental approaches also influence the yield loss results (Emberson et al. 2009). The crop cultivars used by Avnery et al. 2011a, b may have different sensitivity levels as compared to those used in NCLAN and EOTP programmes. Feng and Kobayashi (2009) through meta- analytical studies observed that mean yield loss for soybean and wheat cultivars was ~8 and 10%, respectively at an average O_3 level of ~ 40 ppbv. Mills et al. (2007a) also observed that RYL for wheat at AOT40 of ~23 ppm.h varied from 30–50% depending upon the crop cultivars. Researchers have shown that FACE soybean experiments produced higher yield losses than reported in earlier chamber studies (Long et al. 2005; Morgan et al. 2006).

Based upon MOZRAT-2, Avnery et al. (2011b) predicted the yield losses for soybean, maize and wheat by 2030 according to the two trajectories of O_3 pollution: A2 (pessimistic emission scenario) and B1 (optimistic emission scenario), which represent the upper and lower boundary projections as defined by IPCC (Nakicenovic et al. 2000) (Figs. 4.7 and 4.8). It was estimated that under A2 scenario, relative to year 2000, O_3 induced RYL is estimated to increase by 1.5–10% for wheat, 0.9–10% for soybean and 2.1–3.2% for maize in 2030. Maximum reductions are expected in south Asia (10%) followed by Africa and Middle East (9.4%), eastern Europe (5.8%) and east Asia (5%). For soybean, the yield losses are estimated to be

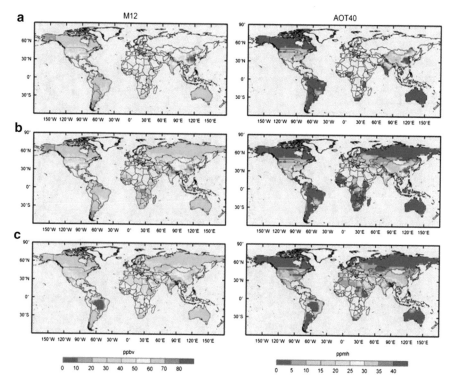

Fig. 4.7 Global distribution of ozone exposure according to M12 and AOT40 metrices under the 2030 B1 scenario During the respective growing seasons of (**a**) soybean (**b**) maize and (**c**) wheat (Source: Avnery et al. 2011b)

the greatest for east Asia (5%), south Asia (11%), EU- 25 (7%) and Africa and Middle East (6.2%). RYL for maize is projected to be influenced mostly in south and east Asia (6.8 and 4.7% respectively) along with a 3% increase for EU- 25 and eastern Europe (Table 4.2).

Under the B1 scenario, global RYL is predicted to be 4–17% for wheat, 10–15% for soybean and 2.5–6% for maize (Table 4.2). Wheat yield by the year 2030 is expected to decrease by ~ 4.1% in south Asia, with the developing regions (Latin America, East Asia, Africa and Middle East) expected to show an additional loss of ~ 1–2%, whereas North America and EU-25, are projected to experience yield gains of 1.7 and 0.8%, respectively as compared to year 2000 (Avnery et al. 2011b). For soybean, additional yield losses are predicted mostly in east and south Asia (8.2 and 4.9%, respectively) followed by Latin America, Africa and Middle East (2%). For EU-25 and North America, soybean yield gains of 2–3% are predicted. Under B1 scenario, South and East Asia are expected to suffer additional maize loses of 3.5 and 2.2%, respectively (Avnery et al. 2011b).

Avnery et al. (2011b) calculated CPL in A2 scenario to be 29–178 Mt. for wheat (a decrease of 9–85 Mt. from the year 2000), 25–53 Mt. for maize (a decrease of 13–20 Mt) and 28–37 Mt. for soybean (a decrease of 11–13 Mt. from the year 2000). Under this scenario, global EL due to O$_3$ induced yield losses was estimated

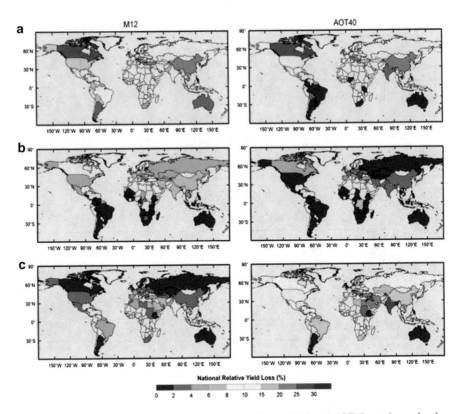

Fig. 4.8 Global distribution of ozone exposure according to M12 and AOT40 metrics under the 2030 A2 scenario during the respective growing seasons of (**a**) soybean (**b**) maize and (**c**) wheat (Source: Avnery et al. 2011b)

to be \$ 17–35 billion annually, which is \$ 6–17 billion more than the year 2000 (Avnery et al. 2011b). This study predicted maximum EL for China (\$ 5.6 billion, an increased loss of \$ 2.6 billion from 2000), followed by India (\$5.2 billion, \$ 2.7 billion more than 2000) and USA (\$ 4.2 billion, \$ 1.1 billion more than 2000) (Avnery et al. 2011b). Under B1 scenario, Avnery et al. (2011b) estimated global CPL for the year 2030 to be 21–106 Mt. for wheat (a decrease in production of 0.8–13 Mt. from the year 2000), 14–35 Mt. for maize (a decrease of 1.7–2.9 Mt. from the year 2000) and 17–27 Mt. for soybean (a decrease of 1.5–1.9 Mt. from the year 2000). This scenario suggests that south Asia will experience greatest additional wheat CPL but the magnitude is greatly reduced as compared to A2 scenario (6.4 Mt. in B1 scenario as compared to 19 Mt. in A2 scenario). Similarly, CPL for maize and soybean is projected to be maximum for east Asia (2–3 Mt. for each crop). However, production gains of 5–6 Mt. of soybean, maize and wheat are anticipated in North America due to reduction in O₃ precursors as per B1 scenario (Avnery et al. 2011b). The CPL for developing countries under the B1 scenario is anticipated to increase in the year 2030 owing to increased O₃ pollution, however, the magnitude

Table 4.2 Estimated Relative Yield Loss (RYL) (%) by year 2030 due to ozone exposure under A2 and B1 scenario

	World		EU-25		USSR & E. Europe		North America		Latin America		Africa & Middle East		E. Asia		S. Asia		ASEAN & Australia	
	A2	B1	A2	B1	A2	B1	A2	B1	A2	B1	A2	B1	A2	B1	A2	B1	A2	B1
WHEAT																		
AOT40	25.8	17.2	16.9	10.4	21.5	11.4	14.5	8.2	12.6	8.1	35.5	21.4	25.7	19.7	44.4	33.8	1.3	1.0
M7	5.4	4.0	4.5	3.4	4.0	2.4	3.1	2.0	3.0	2.6	9.4	6.4	3.8	3.1	11.2	9.2	0	0
Mean	15.6	10.6	10.7	6.9	12.7	6.9	8.8	5.1	7.8	5.4	22.4	13.9	14.7	11.4	27.8	21.5	0.6	0.5
MAIZE																		
AOT40	4.4	2.5	5.9	2.9	5.1	2.2	3.4	1.6	1.2	0.4	1.6	0.8	7.9	5.8	8.9	6.3	2.3	1.2
M12	8.7	6.0	11.0	7.2	9.7	6.4	7.2	4.4	4.6	3.3	5.2	3.6	13.3	10.3	16	12.0	5.9	4.0
Mean	6.5	4.3	8.5	5.0	7.4	4.3	5.3	3.0	2.9	1.9	3.4	2.2	10.6	8.0	12.5	9.1	4.1	2.6
SOYBEAN																		
AOT40	19.0	9.5	32.8	20.4	–	–	15.7	9.8	3.2	1.7	7.8	3.0	40.6	31.5	15.6	8.6	1.4	0.1
M12	14.8	14.6	32.4	25.3	–	–	19.9	14.6	11.9	9.4	16.6	13.3	35.4	30.5	22.0	17.6	9.1	5.7
Mean	16.4	12.1	32.6	22.9	–	–	17.8	12.2	7.5	5.5	12.2	8.2	38.0	31.0	18.8	13.1	5.3	2.9

Source: Avnery et al. (2011b)

of losses is not as severe as predicted under A2 scenario (Avnery et al. 2011b). Global CPL for all the three crops was in the range of 84–137 Mt., approximately 10% greater than the estimate for year 2000. The global EL in the B1 scenario was $ 12–21 billion, which was $ 1–3 billion more from the year 2000 (Avnery et al. 2011b).

O_3 is generally higher in northern hemisphere, however, O_3 exposure during wheat growing season in Brazil and Democratic Republic of Congo in southern hemisphere is expected to increase in near future (Avnery et al. 2011b). Avnery et al. (2011a) have reported that soybean and maize growing seasons experience higher O_3 concentrations as they coincide with the biomass burning seasons of these nations (Avnery et al. 2011a).

Danh et al. (2016) assessed the rice yield loss due to exposure to O_3 pollution in southern Vietnam using O_3 stimulation through a 3D chemistry transport model [AMx] (ENVIRON 2011). This study used three rice crop cycles in a year, each lasting 105 days. These crops are the winter spring crop, the summer autumn crop and the main rice crop grown from mid July to the end of October (Danh et al. 2016). This study used two O_3 matrices; exposure based metric (AOT40 and M7) and flux based phytotoxic O_3 dose (POD_{10}) metric to evaluate O_3 induced relative rice production loss (RPL) in the eastern regions of southern Vietnam. M12 is considered to be a better metrics than M7 but no dose response relationship based on M12 is reported (Sinha et al. 2015). AOT40 is aimed at predicting risks by O_3 exposure rather than validating actual O_3 related cause effect relationships in plants. POD_{10} directly links the O_3 levels to plant uptake and is generally recommended for crop los assessment (Yamaguchi et al. 2014). AOT40 and M7 O_3 exposure indices have been used for quite sometimes, but POD_{10} is a recent approach developed for rice based on experimental studies for Japanese rice cultivar 'Koshihikari' (Yamaguchi et al. 2014). AOT40 and M7 values followed a similar trend, highest in the first crop, followed by the third crop and lowest in the second crop (Table 4.3) (Danh et al. 2016). During the study, the AOT40 approximately reached or exceeded the critical level of 3000 ppb.h for protection of agricultural crops (WHO 2000; LRTAP 2004) as observed in Table 4.3.

RPL reported by Danh et al. (2016) was consistent with the calculated O_3 i.e. minimum loss was recorded in first crop followed by third crop and lowest for the second crop (Table 4.3). The total loss in the domain based on the highest estimates among the three metrics reached approximately 25,800, 21,500 and 6800 Mt., which accounted for 5.7, 3.8 and 1.7% of total rice production respectively, for the first, third and second crop (Danh et al. 2016). In this study, RYL (%) was lowest for the second crop i.e. 1.7% based on AOT40, 0.51% based on POD_{10} and 0.36% based on M7 (Table 4.3). The discrepancies between AOT40 and M7 can be attributed to the duration of daytime hours which may be well extended beyond 7 h underestimating the O_3 effects (Danh et al. 2016). Van Dingenen (2009) indicated that O_3 induced reduction in rice yield in China based on AOT40 and M7 was 3.9 and 3.1%, respectively while the corresponding values for India was 8.3 and 5.7%, respectively. Predictions by Van Dingenen (2009) also supported the observations made by Danh et al. (2016) that M7 provided lower estimated yield losses compared to AOT40.

Table 4.3 O_3 concentration (as per different metrics) Rice Production Loss (RPL) and Relative Yield Loss (RYL) of three rice growing seasons (Danh et al. 2016)

	I Crop (Mid November – End February)			II Crop (Mid March- End June)			III Crop (Mid July – End October)		
	AOT40 (ppb.h)	M7 (ppb)	POD_{10} (mmolm^{-2})	AOT40 (ppb.h)	M7 (ppb)	POD_{10} (mmolm^{-2})	AOT40 (ppb.h)	M7 (ppb)	POD_{10} (mmolm^{-2})
O_3 Concentration	284 ± 81–4183 ± 1055	23.5 ± 1.3–32.3 ± 2.5	0.006 ± 0.06–0.17 ± 0.05	3.2 ± 2.0–1534 ± 574	23.0 ± 1.6–31.4 ± 2.0	0.004 ± 0.001	233.0 ± 80–1781 ± 355	24.0 ± 1.3–32.0 ± 1.4	0.022 ± 0.02–0.21 ± 0.08
AP (Mt)	398, 600			391, 000			533, 100		
RPL (Mt)	16, 967	1819	25, 826	6839	1456	2055	8707	1857	21, 563
RYL (%)	4	0.46	5.7	1.7	0.36	0.51	1.6	0.35	3.8

AP Actual Rice Production, *RPL* Rice Production Loss, *RYL* Relative Yield Loss

The higher RYL estimated using POD_{10}, as observed by Danh et al. (2016) is consistent with other studies which showed that flux based metric showed higher RYL estimates compared to AOT40 (Simpson et al. 2007; Pleijel et al. 2004; Danielsson 2003).

Feng et al. (2015) evaluated the impact of O_3 on wheat and rice through the data obtained from five sites in China equipped with OTCs or O_3- FACE (Free Air O_3 Concentration Enrichment) system and observed that on the basis of exposure concentration and stomatal O_3 - flux response relationship obtained, the current and future O_3 levels induce wheat yield loss by 6.4–14.9% and 14.8–23%, respectively. Yield losses of 4.7, 10.5 and 58.6% respectively in wheat and 7.3, 8.2 and 26.1%, respectively in rice were recorded at ambient, 50 and 100 ppb O_3 as compared to charcoal filtered air (Feng et al. 2003). Wang et al. (2012b) recorded reductions in the range of 8.5–58% and 40–73%, respectively for wheat, and 10–34% and 16–43%, respectively for rice at two O_3 treatments, O_3–1 (75–100 ppb) and O_3–2 (150–200 ppb) as compared to charcoal filtered (CF) air. Zhu et al. (2011) also recorded a reduction of 20% in wheat yield at O_3 concentration of 45.7 ppb, whereas Shi et al. (2009) reported an average loss of 12%, across four rice cultivars. Meta analysis based on 42 wheat experiments performed in Asia showed significant negative effect on grain yield with a decline of 4.7% per 10 ppb O_3 concentration (Borberg et al. 2015).

Soybean (*Glycine max* L.) ranks among the most O_3 sensitive agricultural crops (Mills et al. 2007b) and the fifth most significant crop in terms of global production (FAO 2012). It has been estimated that ground level O_3 pollution causes significant yield reductions in soybean (Van Dingenen 2009; Avnery et al. 2011a, b). In addition to the NCLAN programme in USA, O_3 related soybean yield loss studies have been conducted through Urbane- Champaign and USDA Agricultural Research Services SoyFACE facility at University of Illinios (Long et al. 2005; Betzelberger et al. 2010, 2012; Gillespie et al. 2012). Dose response functions for soybean were developed by Lesser et al. (1990) which was based upon NCLAN data set and by Mills et al. (2007a) which updated the earlier dose response function by combining NCLAN data with other dataset collected from recent soybean studies in US. The dose response function developed by Lesser et al. (1990) was used by Wang and Mauzerall (2004) in their soybean yield loss assessment studies for East Asia and by Van Dingenen (2009) for global assessment of soybean yield losses. Avnery et al. (2011a, b) utilized the dose response function estimated by Mills et al. (2007a) for evaluating soybean yield loss on global scale. A meta analytical study from 53 peer reviewed studies has shown that seed yield in soybean decreased by 24% at an average chronic O_3 exposure of 70 ppb (Morgan et al. 2003). This study showed that at low O_3, a reduction in seed yield corresponded to a decrease in leaf photosynthesis, however, at high O_3 concentration, both leaf photosynthesis and leaf area were responsible for reduced seed yield (Morgan et al. 2006). It was observed that an increase of 23% in ambient O_3 concentration from 56 to 69 ppb, using FACE technology for O_3 fumigation led to a 20% decrease in the seed yield over two growing seasons (Morgan et al. 2006). Betzelberger et al. (2010) studied the response of 10 soybean cultivars grown at elevated O_3 concentration from germination to maturity in SoyFACE programme in 2007 and 6 cultivars in 2008 and based upon the results predicted that doubling background O_3 decreased soybean yield by 17%, however, the variation in response among the cultivars and years ranged between 8–37%. In an another experiment, Betzelberger et al. (2012) used 7 soybean cultivars exposed to O_3 concentration varying from 38–120 ppb for two experimental years (2009 and 2010) and observed a negative correlation between seed yield and O_3 concentration.

Osborne et al. (2016) collated O_3 exposure yield data from 49 soybean cultivars from 28 experimental studies, published between 1982 and 2014 to develop updated dose response fumigation for soybean. Yield reductions varied with different cultivars and ranged from 13.3% for least sensitive cultivar (of the study) "Hodgson" to 37.9% for the most sensitive cultivar "Pusa 9814" at 7 h mean O_3 concentration (M7) of 55 ppb (Osborne et al. 2016). This study indicates that soybean sensitivity increased by an average of 32.5% between 1960 and 2000 suggesting that modern soybean cultivars have a greater sensitivity than older ones. An analysis of O_3 sensitivity of USA cultivars showed that RYL for soybean was estimated to be 14.1% for cultivars released in 1960 compared to 19.3% for cultivars released in 2000. However, the sensitivity- time relationship for Indian cultivars indicate a greater change in sensitivity over time, with RYL increasing from 13.1% in 1960 to 22.6% in 2008. Results suggest that three most sensitive soybean cultivars, PK472, Pusa 9712 and Pusa 9814 were from India, which is prone to higher O_3 concentration in near future (Osborne et al. 2016). Since Emberson et al. (2009) have already proved higher sensitivity of Asian cultivars of soybean compared to North American cultivars; the study by Osborne et al. (2016) confirms these findings.

In addition to the extensive yield loss assessment programmes like NCLAN and EOTP, modeling and meta analytical studies, several individual experiments have also authenticated O_3 induced significant yield reductions in agricultural crops (Table 4.4). As predicted by Deneter et al. (2006), regions of south and east Asia will experience maximum increments in the concentration of background O_3 and its precursors, yield loss assessment studies are important in south Asian regions. Further, this area not only supports the maximum population load, it also holds a major proportion of under nourished population (FAO 2015). These facts point out the importance of yield loss assessment studies in countries like India, China and Pakistan. Last few decades have seen extensive experimental studies conducted in different parts of South and East Asia indicating significant O_3 induced yield reductions in important staple crops of this area (Ismail et al. 2015; Zhang et al. 2014; Yamaguchi et al. 2014; Li et al. 2016). Figure 4.9 shows important locations in the south and east Asia where O_3 induced yield loss assessment studies have been conducted. Along with the filtration and fumigation experimental techniques, antiozonant EDU has also been frequently used in assessing O_3 induced yield losses. The yield response of the EDU treated plants clearly indicate towards the partial protection provided by EDU to O_3 stressed plants (Feng et al. 2015; Yuan et al. 2015). Table 4.5 pools in a few EDU studies which show significant positive effects of EDU treatment on O_3 exposed plants.

Table 4.4 Variation in yield components of different crops upon ozone exposure

Species	Cultivar	Experimental setup	O₃ concentration (ppb)	Yield reduction (%)	Reference
1. *Triticum aestivum* L.	Albis	OTCs	1989: 35.7 49.5 62.2 1990: 38.7 55.6 71.4 (7 h mean)	4.7 18.9 29.3 6.9 14.7 32.1	Fuhrer et al. (1992)
2.	Taylor's Horticultural Lingua di Fuoco Saluggia	OTCs	50 (7 h mean)	31.5 28.6 30.8	Schenone and Lorenzini (1992)
3.	Promessa	OTCs	Ambient +50 ppb (4hd⁻¹, 4 d week⁻¹) Ambient +25 ppb (6 h d⁻¹, 5d week⁻¹) Ambient +50 ppb (3 h d⁻¹, 5d week⁻¹)	53.5 3.2 16.8	Finnan et al. (1996)
4.	Riband	Unenclosed chamber system	81 (7 h mean)	13	Ollerenshaw and Lyons (1999)
5.	Nandu	Closed fumigation chamber	65 110 (12 h mean)	12.1 21.2	Meyer et al. (2000)
6	HD 2329	Ambient air	10–15.4 (6 h mean)	0.5–25.5	Agrawal et al. (2003)
7.	–	OTCs	200 ppb 100 ppb	80.5 58.6	Feng et al. (2003)

No.	Cultivar	Method	Ozone concentration	Yield	Reference
8.	Inqlab- 91 Punjab- 96 Pasban- 90	OTCs	72 (8 h mean)	18.0 39.0 43.0	Wahid (2006)
9.	–	OTCs	36.4–48 (8 h mean)	20.7	Rai et al. (2007)
10	Giza- 68	Ambient air	77–166	9–46	Ali et al. (2008)
11	HUW 510 Sonalika	OTCs	45.3 50.4 55.6 (12 h mean) 45.3 50.4 55.6 (12 h mean)	20.0 37.0 46.0 11.0 25.0 38.5	Sarkar and Agrawal (2010)
12	Sufi Bijoy	Commercial type greenhouse	60 ppb 100 ppb 60 ppb 100 ppb	11.5 44.5 33.2 45.6	Akhtar et al. (2010a)
13	HUW-37 K-9107	OTCs	Ambient +10 ppb (4 h)	39 (I year) 12.4 (II year) 40.8 (I year) 14 (II year)	Mishra et al. (2013)
14	PBW-343 M 533	OTCs	53.2 (12 h mean)	19.0 18.8	Rai and Agrawal (2014)

(continued)

Table 4.4 (continued)

	Species	Cultivar	Experimental setup	O_3 concentration (ppb)	Yield reduction (%)	Reference
15		PBW 343 HD 2936	OTCs	59.2 65.35 59.2 65.35 (7 h mean)	15.9 18.4 10.8 11.2	Tomer et al. (2015)
16		FH 8203 FH 7096	OTCs	4612 ppb.h (AOT40)	47.1 56.0	Adrees et al. (2016)
17	*Oryza sativaL.*	IRRI-6 Basmati- 385	OTCs	35.6 (6 h mean)	37 42	Wahid et al. (1995)
18		MR 84 MR- 185	OTCs	32.5 (8 h mean)	3.4 6.3	Ishii et al. (2004)
19		Chainat 1 Suphanburi 1 Suphanburi 60 Suphanburi 90 Klongluang 1 Pathumthani 1 Gorkor 15 Khowdokmai 105	Closed top chambers	24.9	15.9 12.3 5.6 8.2 5.7 16.6 8.2 2.7	Ariyaphanphitak (2004)
20		Klongluang 1 Pathumthani 1 Gorkor 15 Khowdokmali 105	Closed chambers	150 (7 h d^{-1})	30–78	Ariyaphanphitak et al. (2005)
21		–	OTCs	62	24	Ainsworth (2008)
22		–	OTCs	73–77	17 (I year) 14 (II year)	Reid and Fiscus (2008)
23		Saurabh 950 NDR 97	OTCs	30.5–45.4	11.5 16.0	Rai and Agrawal (2008)

24	Shanyou 63 Liangyoupeijiu	O₃- Face	56 59	17.5 15.0	Shi et al. (2009)
25	SY63 WYJ3	Open air- O₃ release system	1.5 x ambient (7 h mean of ambient = 13.8–74.2)	20.7 6.3	Pang et al. (2009)
26	BR11 BR14 BR28 BR29	Commercial type greenhouse	60 ppb (7 h) 100 ppb 60 ppb 100 ppb 60 ppb 100 ppb 60 ppb 100 ppb	17.5 53.2 6.2 23.5 6.2 41.2 9.7 28.3	Akhtar et al. (2010b) Akhtar et al. (2010b)
27	MR- 263 MR- 219 MR- 84 Mahsuri	Closed top chambers	60 (8hd⁻¹, 3d week⁻¹) 120 (8hd⁻¹, 3d week⁻¹) 60 (8hd⁻¹, 3d week⁻¹) 120 (8hd⁻¹, 3d week⁻¹) 60 (8hd⁻¹, 3d week⁻¹) 120 (8hd⁻¹, 3d week⁻¹) 60 (8hd⁻¹, 3d week⁻¹) 120 (8hd⁻¹, 3d week⁻¹)	10.0 13.02 13.0 14.47 6.0 3.0 5.0 8.0	Ismail et al. (2015)

(continued)

Table 4.4 (continued)

	Species	Cultivar	Experimental setup	O_3 concentration (ppb)	Yield reduction (%)	Reference
28		Malviya Dhan 36 Shivani	OTCs	Ambient	12.2	Sarkar et al. (2015)
				Ambient +10 ppb (4 h d^{-1})	19.7	
					28.8	
				Ambient +20 ppb (4 h d^{-1})	13.1	
					28.6	
				Ambient	43.0	
				Ambient +10 ppb (4 h d^{-1})		
				Ambient +20 ppb (4 h d^{-1})		
29	*Zea mays* L.	HQPM1 DHM117	OTCs	Ambient	4.0	Singh et al. (2014)
				Ambient +15 ppb (5 h)	7.2	
				Ambient +30 ppb (5 h)	10.1	
				Ambient	5.5	
				Ambient +15 ppb (5 h)	9.5	
				Ambient +30 ppb (5 h)	13.8	
30	*Glycine max* L.	Merr. Davis	OTCs	104 (7 h d^{-1})	50	Unsworth et al. (1984)
31		Forrest Essex		62.4 (7 h d^{-1}, 5d week^{-1})	32 / 10	Chernikova et al. (2000)
32		Merr	OTCs	75	23.9 (field)	Booker and Fiscus (2005)
				67	27.2 (pots)	
					38.9 (field)	
					41.0 (pots)	
33			OTCs	60–70	47	Bou Jaoude et al. (2008)
34		PK 472 Bragg	OTCs	70 (4 h d^{-1})	20	Singh et al. (2010)
				100 (4 h d^{-1})	33.6	
				70 (4 h d^{-1})	12	
				100 (4 h d^{-1})	30	

35	Chiang Mai 60 Sorjor 5 Srisumrong 1	OTCs	62 (7 h d^{-1})	36 28 32	Thanacharoenchanaphas and Rugchati (2009)
36	A3127 Clark Dwight Holt HS93–4118 IA- 3010 LN97–15076 Loda NE3399 Pana	Soy- FACE	82.5 (I year) 61.3 (II year) (8 h mean)	21 25.5 17.6 (2 year average) 17.6 14.6 (2 year average) 26.5 (2 year average) 17.4 (2 year average) 11.8 (2 year average) 27.4 15.2 (2 year average)	Betzelberger et al. (2010)
37	HF25 HF35 HF55 HN35 HN37 HN65 SN22 SN26 SN31	OTCs	50.5–58.9	39 33 32 46 39 42 44 42 39	Zhang et al. (2014)

(continued)

Table 4.4 (continued)

	Species	Cultivar	Experimental setup	O_3 concentration (ppb)	Yield reduction (%)	Reference
38	*Phaseolus vulgaris* L.	S 156 R 123 R 331	Environmentally controlled field chambers	60	77 19 35	Flowers et al. (2007)
39		Giza 3	Open field	77–166	20–45	Ali et al. (2008)
40		Borlotto Nano Lingua di Fuoco	OTCs	4765 ppb.h (AOT40)	40.6	Gerosa et al. (2009)
41		S 156 R 123	OTCs	80 (9 h d⁻¹, 40 days)	56 31	Scheepers et al. (2013)
42	*Vigna radiate* L.	Malviya Jyoti	Ambient air	9.0–58.5 (6 h mean)	34.3–73.4	Agrawal et al. (2003)
43		Malviya Jyoti		9.7–58.5	22–79	Agrawal et al. (2006)
44		M-28 M- 6601	OTCs	41–73	47.1 51.1	Ahmed (2009)
45		Nm 2006 Nm 2011	OTCS	3110 ppb.h (AOT40)	47.5 56.0	Adrees et al. (2016)
46	*Linum usitatissimum* L.	Padmini T-397	OTCs	Ambient +10 ppb (3 h)	40.5 42.8	Tripathi and Agrawal (2013)
47	*Vicia faba* L.	Lara	Open field	77–166	13–33	Ali et al. (2008)
48	*Pisum sativum* L.	Perfection	Open field	77–166	3–30	Ali et al. (2008)
49		Climax Peas 09	OTCs	5120 ppb.h (AOT40)	48 52	Adrees et al. (2016)
50	*Brassica campestris* L.	Pusa Jaikishan	Ambient air	10–15.4	5.9–29.7	Agrawal et al. (2003)
51		Kranti	OTCs	41.6–54.2	16.4	Singh et al. (2009)
52		Sanjukta Vardan	OTCs	Ambient +10 ppb (3 h)	12.5 33.4	Tripathi and Agrawal (2012)
53	*Brassica napus* L.	Eurol	Closed chamber	150 (7 h d⁻¹)	14.0	Ollerenshaw et al. (1999)

54	Gossypium hirsutumL.	Giza 65	Closed chamber	70 (10 h d⁻¹, 2 weeks)	23.0	Hassan and Tewfik (2006)
55		Romanos Allegria	Close and environment controlled chamber	100 (7 h d⁻¹)	60.5 51.5	Zouzoulas et al. (2009)
56	Trifolium repensL.	Regal- NC-R NC-S	OTCs	49–70 (7 h mean)	20–60 (in comparison to NC-S)	Fumagalli et al. (2003)
57	Trifolium alexandriumL.	Bundel Wardan JHB-146 Fahli Saidi Mescavi	OTCs	Ambient +10 ppb (6 h mean)	13.5 18.2 9.1 4.9 6.5 4.4	Chaudhary and Agrawal (2013)
58	Trifolium cherleri	–	OTCs	Ambient +40 ppb	32.0	Sanz et al. (2014)
59	Solanum tuberosumL.	Hela	Closed exposure chamber	65 (24 h mean) 110 (4 h d⁻¹)	24 11	Köllner and Krause (2000)
60		Bintje	OTCs	43 (12 h mean)	11.4	Persson et al. (2003)
61		Kardan Bintje	OTCs	31 57 31 57	27.8 11.4 20.0 7.7	Piikki et al. (2004)
62		Desiree	Green house	25.8 42.5 (12 h mean)	53 65	Calvo Er al. (2009)
63	Daucus carotaL.	Pusa Kesar	OTCs	38.4 (8 h mean)	45.3	Tiwari et al. (2006)
64	Raphnus sativaL.	Pusa reshmi	OTCs	36–52 (8 h mean)	29.5	Tiwari and Agrawal (2010)

(continued)

Table 4.4 (continued)

	Species	Cultivar	Experimental setup	O$_3$ concentration (ppb)	Yield reduction (%)	Reference
65	*Beta vulgaris* L.	Patriot	OTCs	62 (8 h mean)	14	De Temmerman et al. (2007)
66		Allgreen	OTCs	38.15 (8 h mean)	23.9	Tiwari et al. (2010)
				51 (8 h mean)	28.6	
67		Allgreen	OTCs	53 (8 h mean)	25.9	Kumari et al. (2013)
68	*Lycopersicon esculentus* L.	Pusa ruby	Closed top dynamic chamber	75	35	Mina et al. (2010)
				150	68	
				(2 h d^{-1}, 12 days)		
69		Triton	Controlled environmental chamber	75 (7 h d^{-1})	19.4	Gillespie et al. (2015)
70	*Solanum lyopersicon* Mill.	Look Ver	Closed chambers	200	10	Thwe et al. (2014)
				350	17	
				500	22	
				(4 h d^{-1} for 6 days)		
71	*Fragaria ananassa* Duch.	Bogota Elsanta	OTCs	74 (8 h d^{-1})	40.7 NS	Drogoudi and Ashmore (2002)
72	*Capsicum annuum* L..	–	OTCs	78 (8 h d^{-1} for 75 days)	29	Taia et al. (2013)

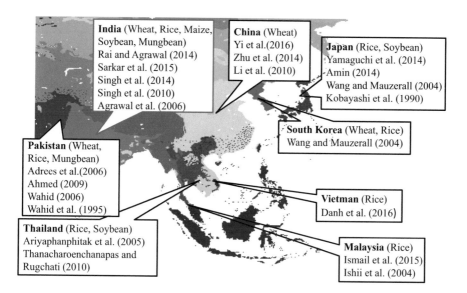

Fig. 4.9 Ozone induced yield reduction studies done in South and South East Asia

Several experiments using filtration/fumigation approach have given a picture of serious consequences of O_3 induced yield reductions on global food security, especially in the regions of south and south-east Asia (Sarkar et al. 2015; Tomer et al. 2015; Rai and Agrawal 2014; Mishra et al. 2013; Akhtar et al. 2010a, b). In addition to crop loss assessment experiments conducted for wheat, rice and soybean, yield loss for other crops like mungbean and vegetables like palak have also been studied (Tiwari and Agrawal 2011; Agrawal et al. 2006, 2003). In a transect study conducted at Pakistan, two pea cultivars *Pisum sativum* L. var. Climax and Peas 09 showed significant reductions of 48 and 52%, respectively at O_3 exposure of 5120 ppm.h (AOT40) (Adrees et al. 2016). In the same experiment, two cultivars each of wheat (*Triticum aestivum* L var. FH 8203 and FH 7096) and mungbean (*Vigna radiate* L. var. NM 2006 and NM 2011) showed yield reductions of 47and 56%, respectively when O_3 concentration was 4612 ppm.h during the growing season of wheat and 3110 ppm.h during the growing season of mungbean (Adrees et al. 2016). Ahmed (2009) showed reductions of 47.06 and 51.12% in seed yield of two mungbean varieties M-28 and 6601, when 8 h mean O_3 concentration ranged between 41 to 73 ppb. In another transect study, Agrawal et al. (2006) reported significant reductions in the yield of mungbean (*Vigna radiate* L. var. Malviya Jyoti) along a transect gradient with 6 h O_3 concentration ranging from 9.7 to 58.5 ppb. The yield reduction was found to be maximum (77%) at the site where O_3 concentration was maximum (58.5 ppb). In yet another transect study, Agrawal et al. (2003) reported yield reductions of 73, 25, 20 and 40.4% in *Vigna radiate* L. var. Maliya Jyoti, *Triticum aestivum* L. var. HD2329, *Brassica compestris* L. var. Pusa Jaikishan and *Beta vulgaris* L. var. Allgreen, respectively, at a site experiencing maximum concentration of O_3 (58.50 ppb in summer and 15.33 ppb in winter) as compared to

Table 4.5 Yield response of ozone exposed plants upon EDU treatment

Plant	Species	O₃ concentration	EDU dose	% change in yield	Reference
Triticum aestivum L.	HUW 234	35.3–54.2 ppb (Mean monthly)	400 ppm	(+) 8.6	Singh and Agrawal (2009)
	HUW 468			(+) 25.6	
	HUW 510			(+) 20.4	
	PBW 343			(+) 24	
	Sonalika			(+) 1.9	
	Yangmei 185	–	150 ppm	(+) 3.4	Wang et al. (2007)
			300 ppm	(+) 12.7	
			450 ppm	(+) 7.1	
	Malviya 234	–	150 ppm	(+) 24.8	Tiwari et al. (2005)
			300 ppm	(+) 66.9	
			450 ppm	(+) 66.8	
	Malviya 533	–	150 ppm	(+) 18.8	Tiwari et al. (2005)
			300 ppm	(+) 19.1	
			450 ppm	(+) 20.5	
	HUW 234	29.2 ppb (Average)	500 ppm	(+) 27	Agrawal et al. (2004)
	HUW 468			(+) 22	
	HUW 2329			(+) 36.3	
Oryza Sativa L.	Jiahua 2	–	150 ppm	(+) 0.15 (ns)	Wang et al. (2007)
			300 ppm	(+) 3.8 (ns)	
			450 ppm	(+) 3.4 (ns)	
Phaseolus vulgaris L.	S156	41–59 ppb (Average hourly)	300 ppm	(+) 71	Elagoz and Manning (2005)
	R123			(−) 38	
	BBL- 290			(−) 5 ns	
	BBL- 274			(+) 3 ns (Seed dry weight)	
	cv. Lit.	14.98 ppb 31.56 ppb	150 ppm	(+) 65	Brunschon-Harti et al. (1995)
	cv. Lit			(+) 31	
Vigna radiate L.	Malviya Janpriya	52.9–64.5 ppb (mean monthly)	400 ppm	(+) 32.28	Singh et al. (2010b)
	Malviya Jyoti	32.64–35.19 ppb (mean monthly)	500 ppm	(+) 32.2	Agrawal et al. (2005)
	cv. Pratikshya	29.3–39.1 ppb (mean monthly)	400 ppm	(+) 24	Kharel and Amgain (2010)
	Azad- 1	41.3–59.9 ppb	400 ppm	(+) 44.4	Singh and Agrawal (2011b)
	BHU- 1			(+) 40.9	

(continued)

Table 4.5 (continued)

Plant	Species	O₃ concentration	EDU dose	% change in yield	Reference
Glycine max L.	Narc- I	40 ppb	400 ppm	(+) 96.96	Wahid et al. (2001)
		48 ppb		(+) 94.3	
		63 ppb		(+) 112.82	
		75 ppb		(+) 181.81	
		70 ppb		(+) 284.61	
	Pusa 9814	50.5–58.9 ppb (Mean monthly)	400 ppm	(+) 28.2	Singh and Agrawal (2011a)
	Pusa 9712			(+) 29.8	
Trifolium alexandrinum L.	cv. Messkawy	88 ppb (8 h average)	50 ppm	(+) 14.28	Hassan et al. (2007)
			100 ppm	(+) 62.85	
			150 ppm	(+) 74.28	
			200 ppm	(+) 74.28	
Trifolium repens L.	cv Menna	3000 ppb.h (AOT40)	150 ppm	(−)17.3 (EDU Biomass ratio)	Ball et al. (1998)
	cv. Vardan	30.3–46.6 ppb (12 h mean)	300 ppm	(+) 46.2	Singh et al. (2010c)
	cv Bundel			(+) 21	
Trifolium subterraneum L.	cv Geraldton	51 ppb (7 h mean)	150 ppm	(+) 13	Tonneijk and van Dijk (1997)
Arachis hypogaea L.	cv P1268661	43 ppb (7 h, 4d)	2 g of 50% EDU	(+) 24	Ensing et al. (1985)
	cv. McRan			NS	
Daucus carota L.	PusaKesar	36.1 ppb (average)	150 ppm	(+) 22.8	Tiwari and Agrawal (2010)
			300 ppm	NS	
			450 ppm	NS	
Beta vulgaris L.	Allgreen	52–73 ppb (mean monthly)	300 ppm	(+) 28.9	Tiwari and Agrawal (2009)
Solanum tuberosum L.	cv. Kara	81 ppb	300 ppm	(+) 35	Hassan (2006)
Raphanus sativus L.	cv. Cherry Belle	36 ppb (7 h mean)	200 ppm	(+) 32	Pleijel et al. (1999)
Raphanus sativus L.	Local Egyptian variety	54.8 ppb (6 h mean)	500 ppm	(+) 33.3	Hassan et al. (1995)
		66.9 ppb (6 h mean)		(+) 36.2	

(continued)

Table 4.5 (continued)

Plant	Species	O₃ concentration	EDU dose	% change in yield	Reference
Brassica rapa L.		54.8 ppb (6 h mean)	500 ppm	NS	Hassan et al. (1995)
		66.9 ppb (6 h mean)		(+) 20.9	
Glycine max L.	cv. JS 335	42 ppb (12 h mean)	400 ppm	(+) 26.7	Rai et al. (2015)
Phaseolus vulgaris L.	cv. S156	71.3 ppb (8 h mean)	450 ppm	(+) 55	Yuan et al. (2015)
	cv. DDW			(+) 46	
	cv. R123			NS	
	cv. NX816			NS	
Brassica compestris L.	cv. Kranti	54.8 ppb (16 ppm.h)	200 ppm	(+) 7	Pandey et al. (2014)
			400 ppm	(+) 17	
	cv. Peela Sona	"	200 ppm	(+) 59	
			400 ppm	(+) 34	

the reference site. Rai and Agrawal (2014) observed a significant reduction of 19 and 18.8%, in the yield (weight of grains plant^{-1}) of two wheat cultivars PBW 343 and M 533, respectively, at 12 h mean O$_3$ concentration of 53.2 ppb. The ability of M533 to sustain seed yield as compared to PBW 343 under similar O$_3$ exposure conditions reflect the greater tolerance of M533 to O$_3$ exposure. Tiwari et al. (2005), in an EDU based experiment also observed M533 to be an O$_3$ tolerant cultivar.

Sarkar et al. (2015) studied the effect of ambient and elevated O$_3$ on two high yielding cultivars of rice (*Oryza sativa* L. var. Malviya Dhan 36 and Shivani) and observed significant reductions in yield (weight of grains plant^{-1}) of the two cultivars. Reductions of 12.2, 19.7 and 28.8% were recorded in yield of Malviya Dhan 36 at ambient, ambient +10 ppb and ambient +20 ppb O$_3$ concentrations, respectively, while Shivani showed reductions of 13.1, 28.6 and 43%, respectively under similar O$_3$ exposure regimes. The results showed that yield reductions in Shivani were higher than Malviya Dhan 36 under similar O$_3$ exposure conditions (Sarkar et al. 2015). The variations in the yield of the two cultivars could be explained by differential partitioning of the photosynthates between grains and above ground biomass. Cultivar Shivani utilized more of its photosynthates for defense and repair mechanisms leading to lesser translocation of assimilate towards reproductive parts, resulting in higher yield reductions (Sarkar et al. 2015). Akhtar et al. (2010a) studied the response of four Bangladeshi rice cultivars (BR11, BR14, BR28 and BR29) in a transect study in which 7 h O$_3$ monitored at two sites was 60 and 100 ppb, respectively. Yield reductions (determined by number of panicles plant^{-1} and percentage of filled grains panicle^{-1}) recorded for BR11, BR14, BR28 and BR29 were 17.5, 6.1, 6.2 and 9.65%, respectively at 60 ppb and 53.2, 23.57, 41.2 and 28.3%, respectively at 100 ppb O$_3$ concentration (Akhtar et al. 2010a). An analysis of the growth parameters showed that O_3 exposure resulted in reduced number of tillers plant^{-1}, which led to a lesser number of panicles plant^{-1}. Grain sterility, which is an important factor contributing towards O_3 induced yield reduction (Ishii et al. 2004) was found to increase significantly upon O_3 exposure (Akhtar et al. 2010a). Several other filtration and fumigation studies done around the globe clearly indicated the negative effect of ground level O_3 to crop productivity (Table 4.4).

5 Conclusion

Studies conducted around the globe give a clear cut evidence of the extensive negative impacts of the current ambient O_3 on crops and vegetation. As the O_3 induced plant damage is significant not only in developing but also in developed countries of the world, the problem acquires a global outlook. In spite of the implementation of current legislation, O_3 concentrations are predicted to follow an upward trend in the near future and thus it is likely that the effect of O_3 on crop and vegetation will worsen. Due to the uncontrolled and unplanned emissions of O_3 precursors, the developing countries especially those of south and east Asia are predicted to face a more severe increase in O_3 concentrations. In addition to this, the crop varieties of key staple crops used for cultivation in these regions are more sensitive to O_3 exposure as compared to their respective varieties used in American or European agroecosystems. Current scenario of O_3 pollution hassled to the establishment of several biomonitoring programmes which assist in establishing the severity of the damage done to the plants upon O_3 exposure. Although visible foliar injury forms the basis of the evaluation of O_3 stress in these biomonitoring programmes, presence of foliar injury symptoms do not necessarily imply that there are significant negative effects on growth, yield or reproduction. The biomonitoring programmes have identified sensitive species to be further investigated under controlled conditions or using O_3 protectant, ethylenediurea (EDU). Biomonitoring studies provide relevant information that helps to frame O_3 policy issues related to vegetation protection. Predictions by few modeling studies using simulated O_3 concentrations indicate the global risk the important staple crops may experience due to ground level O_3 pollution in near future. On the basis of the recorded crop loss production, it is clearly evident that O_3 will be an important component in the present climate change scenario and will play a significant role in determining the threats to global food security in near future.

References

Adams RM, Glyer JD, Johnson SL, McCarl BA (1989) A reassessment of theeconomic effects of ozone on United States agriculture. J Air Pollut Control Assoc 39:960–968

Adrees M, Saleem F, Jabeen F, Rizwan M, Ali S, Khalid S, Ibrahim M, Iqbal N, Abbas F (2016) Effects of ambient gaseous pollutants on photosynthesis, growth, yield and grain quality of selected crops grown at different sites varying in pollution levels. Arch Agronom Sci 62(9):34–47

Agathokleous E, Mouzaki-Paxinou A, Saitanis CJ, Paoletti E, ManningWJ (2016a) The first toxicological study of the antiozonant andresearch tool ethylenediurea (EDU) using *Lemna minor* L. bioassay:hints to its mode of action. Environ Pollut 213:996–1006

Agathokleous E, Saitanis CJ, Stamatelopoulos D, Mouzaki-Paxinou A, Paoletti E, Manning WJ (2016b) Olive oil for dressing plantleaves so as to avoid O3 injury. Water Air Soil Pollut 227:282

Agathokleous E, Paoletti E, Saitanis CJ, Manning WJ, Shi C, Koike T (2016c) High doses of ethylenediurea (EDU) are not toxic to willowand act as nitrogen fertilizer. Sci Total Environ 566–567:841–850

Agathokleous E, Paoletti E, Saitanis CJ, Manning WJ, Sugai T, Koike T (2016d) Impacts of ethylenediurea (EDU) soil drench and foliar spray in *Salix sachalinensis* protection against O3 induced injury. Sci Total Environ 573:1053–1062

Agrawal M, Rajput M, Singh RK (2003) Use of ethylenediurea to assess the effects of ambient ozone on *Vigna radiata*. Int J Biotronics 32:35–48

Agrawal SB, Singh A, Rathore D (2004) Assessing the effects of ambient air pollution on growth, biochemical and yield characteristics of three cultivars of wheat (Triticum aestivum L.) with ethylenediurea and ascorbic acid. J Plant Biol 31:165–172

Agrawal SB, Singh A, Rathore D (2005) Role of ethylenediurea (EDU) in assessing impact of ozone on *Vigna radiata* L. plants in a sub urban area of Allahabad (India). Chemosphere 61:218–228

Agrawal M, Singh B, Agrawal SB, Bell JNB, Marshall F (2006) The effect of air pollution on yield and quality of mung bean grown in peri-urban areas of Varanasi. Water Air Soil Pollut 169:239–254

Ahmed S (2009) Effects of air pollution on yield of mungbean in Lahore, Pakistan. Pak J Bot 41:1013–1021

Ainsworth EA, Yendrek CR, Sitch S, Collins WJ, Emberson LD (2012) Theeffects of tropospheric ozone on net primary productivity and implications forclimate change. Annu Rev Plant Biol **63**:637–661

Ainsworth EA (2008) Rice production in changing climate: a meta-analysis of responses to elevated CO2 and elevated ozone concentration. Glob Chang Biol 14:1642–1650

Akhtar N, Yamaguchi M, Inada H, Hoshino D, Kondo T, Izuta T (2010a) Effects of ozone on growth, yield and leaf gas exchange rates of two Bangladeshi cultivars of wheat (Triticum aestivum L.) Environ Pollut 158:1763–1767

Akhtar N, Yamaguchi M, Inada H, Hoshino D, Kondo T, Fukami M, Funada R, Izuta T (2010b) Effects of ozone on growth, yield and leaf gas exchange rates of four Bangladeshi cultivars of rice (Oryza sativa L.) Environ Pollut 158:2970–2976

Ali A, Alfarhan A, Robinson E, Bokhari N, Al-Rasheid K, Al-Quraishy S (2008) Tropospheric ozone effects on the productivity of some crops in central Saudi Arabia. Am J Environ Sci 4:631–637

Ariyaphanphitak W (2004) Effects of ground-level ozone on crop productivity in Thailand. The Joint International Conference on "Sustainable Energy and Environment (SEE)" 1–3 December 2004, Hua Hin, Thailand

Ariyaphanphitak W, Chidthaisong A, Sarobol E, Bashkin VN, Towprayoon S (2005) Effects of elevated ozone concentrations on Thai jasmine rice cultivars (Oryza sativa L.) Water Air Soil Pollut 167:179–200

Aunan K, Berntsen TK, Seip HM (2000) Surface ozone in China and its possible impact on agricultural crop yields. Ambio 29:294–301

Avnery S, Mauzerall DL, Liu JF, Horowitz LW (2011a) Global crop yieldreductions due to surface ozone exposure: 2 year 2030 potential cropproduction losses and economic damage under two scenarios of O3 pollution. Atmos Environ **45**:2297_2309

Avnery S, Mauzerall DL, Liu JF, Horowitz LW (2011b) Global crop yieldreductions due to surface ozone exposure: 1 year 2000 crop production lossesand economic damage. Atmos Environ 45:2284–2296

Bagard M, Le Thiec D, Delacote E, Hasenfratz-Sauder MP, Banvoy J, Gerard J, Dizengremel P, Jolivet Y (2008) Ozoneinducedchanges in photosynthesis and photorespiration ofhybrid poplar in relation to the development stage of the leaves. Physiol Plant **134**:559–574

Ball GR, Benton J, Palmer-Brown D, Fuhrer J, Skarby L, Gimeno BS, Mills G (1998) Identifying factors which modify the effect of ambient ozone on white clover (Trifolium repens) in Europe. Environ Pollut 103:7–16

Betzelberger AM, Gillespie KM, McGrath JM, Koester RP, Nelson RL, Ainsworth EA (2010) Effects of chronic elevated ozone concentration on antioxidant capacity, photosynthesis and seed yield of 10 soybean cultivars. Plant Cell Environ 33:1569–1581

Bermadinger-Stabentheiner E (1996) Influence of altitude, sampling year and needle age class on stress-physiological reactions of spruce needles investigated onan Alpine altitude profile. J Plant Physiol 148:339–344

Betzelberger AM, Yendrek CR, Sun J, Leisner CP, Nelson RL, Ort DR, Ainsworth EA (2012) Ozone exposure response for US soybean cultivars: linear reductions in photosynthetic potential, biomass, and yield. Plant Physiol 160:1827–1839

Black VJ, Black CR, Roberts JA, Stewart CA (2000) Impact of ozone on the reproductive development of plants. New Phytol 147:421–447

Black VJ, Stewart CA, Roberts JA, Black CR (2010) Direct effects of ozone on reproductive development in Plantage major L. populations differing in sensitivity. Environ Exp Bot 69:121–128

Booker FL, Fiscus EL (2005a) The role of ozone flux and antioxidants in the suppression of ozone injury by elevated CO2 in soybean. J Exp Bot 56(418):2139–2151

Booker FL, Fiscus EL (2005b) The role of ozone flux and antioxidants in the suppression of ozone injury by elevated CO in soybean. J Exp Bot 56(418):2139–2151

Bou JM, Katerji N, Mastrorilli M, Rana G (2008) Analysis of the ozone effect on soybean in the Mediterranean region II. The consequences on growth, yield and water use efficiency. Eur J Agron 28:519–525

Broberg MC, Feng ZZ, Xin Y, Pleijel H (2015) Ozone effects on wheat grain quality: a summary. Environ Pollut 197:203–213

Brunschon-Harti S, Fangmeir A, Jager HJ (1995) Influence of ozone and ethylenediurea on growth and yield of bean (Phaseolus vulgaris) in open top chambers. Environ Pollut 90:84–94

Burney J, Ramanathan V (2014) Recent climate and air pollution impacts on Indian agriculture. Proc Natl Acad Sci U S A 111:16319–16324

Calatayud V, Cerveró J, Sanz MJ (2007) Foliar, physiologial and growth responses of four maple species exposed to ozone. Water Air Soil Pollut 185:239–254

Calvete-Sogo H, Elvira S, Sanz J, Gonzalez-Fernandez I, García-Gomez H, Sanchez-Martín L, Alonso R, Bermejo-Bermejo V (2014) Current ozone levels threaten gross primary production and yield of Mediterranean annual pastures and nitrogen modulates the response. Atmos Environ 95:197–206

Calvo E, Calvo I, Jimenez A, Porcuna JL, Sanz MJ (2009) Using manure to compensate ozone-induced yield loss in potato plants cultivated in the east of Spain. Agric Ecosyst Environ 131:185–192

Carnahan JE, Jennce EL, Wat EKW (1978) Prevention of ozone injury in plants by a new protective chemical. Phytopathology 68:1225–1229

Carriero G, Emiliani G, Giovannelli A, Hoshika Y, Manning WJ, Traversi ML, Paoletti E (2015) Effects of long-term ambient ozone exposure on biomass and wood traits in poplar treated with ethylenediurea (EDU). Environ Pollut 206:575–581

Chameides WL, Kasibhatla PS, Yienger J, Levy H (1994) Growth of continental-scale metro-agro-plexes, regional ozonepollution, and world food production. Science 264:74–77. https://doi.org/10.1126/science.264.5155.74

Chaudhary N, Agrawal SB (2013) Intraspecific responses of six Indian clover cultivars under ambient and elevated levels of ozone. Environ Sci Pollut Res 20:5318–5329

Chernikova T, Robinson JM, Lee EH, Mulchi CL (2000) Ozone tolerance and antioxidant enzyme activity in soybean cultivars. Photosynth Res 64:15–26

Collins WJ, Sitch S, Boucher O (2010) How vegetation impacts affect climate metrics for ozone precursors. Geophys Res 115:D23308. https://doi.org/10.1029/2010JD014187

Cooper OR, Parrish DD, Ziemke J, Balashov NV, Cupeiro M, GalballyIE GS et al (2014) Global distribution and trends of tropospheric ozone: an observation-based review. Elem Sci Anth 2:000029

Cotrozzi L, Remorini D, Pellegrini E, Landi M, Massai R, Nali C, Guidi L, Lorenzini G (2016) Variations in physiological and biochemical traits of oak seedlings grown under drought and ozone stress. Physiol Plant 157:69–84

Coulston JW (2011) Modeling ozone bioindicator injury with micro scale and landscape-scale explanatory variables: A logistic regression approach. chapter 6. In: Conkling BL (ed) Forest health monitoring 2007 nationaltechnical report. Gen. Tech. Rep. SRS-XXX. US Department of Agriculture, Forest Service, Southern Research Station, Asheville, p 13

Cuny D, Davranche L, Van Haluwyn C, Plaisance H, Caron B, Malrieu V (2004) Monitoring ozone by usingtobacco, automated network and passive samplers in an industrial area in France. In: Klumpp A, Ansel W, Klumpp G (eds) Urban air pollution, bioindication and environmental awareness. Cuvillier, Go¨ ttingen, pp 97–108

Danh NT, Huy LH, Oanh NTK (2016) Assessment of rice yield loss due to exposure to ozone pollution in Southern Vietnam. Sci Total Environ 566-567:1069–1079

Danielsson H (2003) Exposure, uptake and effects of ozone. Doctoral Thesis, Göteborg University, Department of Applied Environmental Science

Debaje SB (2014) Estimated crop yield losses due to surface ozone exposure and economic damage in India. Environ Sci Pollut Res 21:7329–7338. https://doi.org/10.1007/s11356-014-2657-6

De Temmerman L, Legrand G, Vandermeiren K (2007) Effects of ozone on sugar beet grown in Open-Top Chambers. Eur J Agron 26:1–9

Dentener F, Stevenson D, Ellingsen K, Van Noije T, Schultz M et al (2006) The global atmospheric environment for the next generation. Environ Sci Technol 40:3586–3594

Deneter F, Stevenson D, Ellingsen K, Van Noije T, Schultz M, Amann M, Atherton C, Bell N, Bergmann D, Bey I, Bouwman L, Butler T, Cofala J, Collins B, Drevet J, Doherty R, Eickhout B, Eskes H, Fiore A, Gauss M, Hauglustaine D, Horowitz L, Isaken ISA, Josse B, Lawrence M, Krol M, Lamarque JF, Montanaro V, Muller J-F, Peuch VH, Pitari G, Pyle J, Rast S, Rodriguez J, Sanderson M, Savage NH, Shindell D, Stahan S, Szopa S, Sudo K, Van Dingenen R, Wild O, Zeng G (2006) The global atmospheric environment for the next generation. Environ Sci Technol 40:3586–3594

Drogoudi PD, Ashmore MR (2002) Effects of elevated ozone on yield and carbon allocation in strawberry cultivars differing in developmental stage. Phyton-Annales Rei Botanicae 42:45–53

EANET (2006) Data Report on the Acid Deposition in the East Asian Region 2005. Network Centre of EANET, Japan. Available from: http://www.eanet.cc/

Elagoz V, Manning WJ (2005) Responses of sensitive and tolerant bush beans (Phaseolus vulgaris L.) to ozone in open-top chambers are influenced by phenotypic difference, morphological characteristics and chamber environment. Environ Pollut 135:371–383

Emberson LD, Bu¨ ker P, Ashmore MR, Mills G, Jackson LS, Agrawal M, Atikuzzaman MD, Cinderby S, Engardt M, Jamir C, Kobayashi K, Oanh NTK, Quadir QF, Wahid AA (2009) comparison of North American and Asian exposure–response data for ozone effects on crop yields. Atmos Environ 43:1945–1953

Ensing J, Hofstra G, Roy RC (1985) The impact of ozone on peanut exposed in the laboratory and field. Phytopathology 75:429–432

ENVIRON (2011) User's Guide: Comprehensive Air Quality Model with Extensions (CAMx). ENVIRON I

EPA, Environmental Protection Agency (1996) Air Quality Criteria for Ozone andRelated Photochemical Oxidants. United States Environmental ProtectionAgency, Washington, DC, pp 1-1e1–1-133

EU European Union (2001) Directive 2001/81/EC of theEuropean parliament and the council on national emissionceilings for certain atmospheric pollutants. Off J Euro Commun L 309:22–30

EU European Union (2002) Directive 2002/3/EC of the Europeanparliament and of the council relating to ozone in ambient air. Off J Euro Commun L 67:14–30

FAO (2012) Agricultural production and natural resource use. In: Alexandratos N, Bruinsma J (eds) World Agriculture Towards 2030/2050: The 2012 Revision. ESA Working Paper No. 12-03. FAO, Rome, pp 94–133

FAO (2015) The State of Food Insecurity in the World 2015. Meeting the 2015 international hunger targets: taking stock of uneven progress. FAO, Rome, Italy

Feng ZW, Jin MH, Zhang FZ, Huang YZ (2003) Effects of ground-level ozone (O) pollution on the yields of rice and winter wheat in the Yangtze River Delta. J Environ Sci (China) 15:360–362

Feng Z, Kobayashi K (2009) Assessing the impacts of current and future concentrations of surface ozone on crop yield with meta analysis. Atmos Environ 43:1510–1519

Feng Z, Wang S, Szantoi Z, Chen S, Wang X (2010) Protection of plants fromambient ozone by applications of ethylenediurea (EDU): a meta-analytic review. Environ Pollut 158:3236–3242

Feng Z et al (2014) Evidence of widespread ozone-induced visible injury on plants in Beijing, China. Environ Pollut 193:296–301

Feng Z, Hu E, Wang X, Jiang L, Liu X (2015) Ground-level O_3 pollution and its impacts on food crops in China: A Review. Environ Pollut 199:42–48

Finnan JM, Jones MB, Burke JL (1996) A time- concentration study on the effects of ozone on spring wheat (Triticum aestivum L.). 1. Effects on yield. Agric Ecosyst Environ 57:159–167

Flowers MD, Fiscus EL, Burkey KO, Booker FL, Dubois JJB (2007) Photosynthesis, chlorophyll fluorescence, and yield of snap bean (Phaseolus vulgaris L.) genotypes differing in sensitivity to ozone. Environ Exp Bot 61:190–198

Fuhrer J, Booker FL (2003) Ecological issues related to ozone:agricultural issues. Environ Int 29:141–154

Fuhrer J, Grandjean Grimm A, Tschannen W, Shariat-Madari H (1992) The response of spring wheat (Triticum aestivum L.) to ozone at higher elevations. II. Changes in yield, yield components and grain quality in response to ozone flux. New Phytol 121:211–219

Fumagalli I, Mignanego L, Mills G (2003) Ozone biomonitoring with clover clones: yield loss and carryover effect under high ambient ozone levels in northern Italy. Agric Ecosyst Environ 95:119–128

Gelang J, Pleijel H, Sild E, Danielsson H, Younis S, Selldén G (2000) Rate and duration of grain filling in relation to flagleaf senescence and grain yield in spring wheat (Triticum aestivum) exposed to different concentrations of ozone. Physiol Plant 110:366–375

Gerosa G, Marzuoli R, Rossini M, Panigada C, Meroni M, Colombo R, Faoro F, Iriti M (2009) A flux-based assessment of the effects of ozone on foliar injury, photosynthesis, and yield of bean (Phaseolus vulgaris L. cv. Borlotto Nano Lingua di Fuoco) in open-top chambers. Environ Pollut 157:1727–1736

Gerosa G, Fusaro L, Monga R, Finco A, Fares S, Manes F, Marzuoli R (2015) A flux-based assessment of above and below ground biomass of Holm oak (Quercus ilex L.) seedlings after one season of exposure to high ozone concentrations. Atmos Environ 113: 41–49

Ghude SD, Jena C, Chate DM, Beig G, Pfister GG, Kumar R, Ramanathan V (2014) Reductions in India's crop yield due to ozone. Geophys Res Lett 41:GL060930. https://doi.org/10.1002/2014gl060930

Gillespie KM, Xu FX, Richter KT et al (2012) Greater antioxidant and respiratory metabolism in field-grown soybean exposed to elevated O under both ambient and elevated CO. Plant Cell Environ 35:169–184

Gillespie C, Stabler D, Tallentire E, Goumenaki E, Barnes J (2015) Exposure to environmentally-relevant levels of ozone negatively influence pollen and fruit development. Environ Pollut 206:494–501

Godzik B, Manning WJ (1998) Relative effectiveness of ethylenediurea and constituents amount of urea and phenylurea in EDU, in prevention of ozone injury to tobacco. Environ Pollut 103:1–6

Gottardinia E, Cristoforia A, Cristofolinia F, Nalib C, Pellegrinib E, Bussottic F, Ferretti M (2014) Chlorophyll-related indicators are linked to visible ozone symptoms: evidence from a field study on native Viburnum lantana L. plants innorthern Italy. Ecol Indic 39:65–74

Grantz DA, Yang S (2000) Ozone impacts on allometry and root hydraulic conductance are not mediated by source limitation nor developmental age. J Exp Bot 51:919–927

Hassan IA (2006) Physiological and biochemical responses of potato (Solanum tuberosum L. v. Kara) to ozone and antioxidant enzymes. Ann Appl Biol 148:197–206

Hassan IA, Tewfik I (2006) CO photoassimilation, chlorophyll fluorescence, lipid peroxidation and yield in cotton (Gossypium hirsutum L. cv Giza 65) in response to O. World Rev Sci Technol Sustainable Dev 3(1):70–79

Hassan IA, Ashmore MR, Bell JNB (1995) Effects of ozone on radish and turnip under Egyptian field conditions. Environ Pollut 89:107–114

Hassan IA, Bell JNB, Marshall FM (2007) Effects of air filtration on Egyptian clover (Trifolium alexandrium L. cv. Messkawy) grown in open top chambers in a rural site in Egypt. Res J Biol Sci 2(4):395–402

Hayes F, Jones MLM, Mills G, Ashmore M (2007) Meta-analysis of the relative sensitivity of semi-natural vegetation species to ozone. Environ Pollut 146:754–762. https://doi.org/10.1016/j.envpol.2006.06.011

Heagle AS (1989) Ozone and crop yield. Annu Rev Phytopathol 27:397–423

Heck WW (1989) Assessment of crop losses from air pollutants in the UnitedStates. In: MacKenzie JJ, El-Ashry MT (eds) Air pollution's toll on forests and crops. Yale University Press, New Haven, pp 235–315

Heck WW, Taylor OC, Tingey DT (eds) (1988) Assessment of crop loss from air pollutants. Elsevier Appl. Sci, London

Horowitz LW et al (2007) Observational constraints on the chemistry of isoprenenitrates over the eastern United States. J Geophys Res 112:D12S08. https://doi.org/10.1029/2006JD007747

Hoshika Y, Watanabe M, Inada N, Koike T (2013) Model-basedanalysis of avoidance of ozone stress by stomatal closure in58 siebold's beech (Fagus crenata). Ann Bot 112:1149–1158

Innes JL, Skelly JM, Schaub M (2001) Ozone and broadleaved species. A guide to the identification of ozone-induced foliar injury. Paul Haupt Publishing, Bern/Switzerland. ISBN 3-258-06384-2, p 136

IPCC: Climate Change (2013) In: Stocker TF, Qin D, Plattner G-K, Tignor M, Allen SK, Boschung J, Nauels A, Xia Y, Bex V, Midgley PM (eds) The physical science basis. Contribution of working group I to the fifth assessment report of the intergovernmental panel on climate change. Cambridge University Press, Cambridge/New York

Ishii S, Marshall FM, Bell JNB, Abdullah AM (2004) Impact of ambient air pollution on locally grown rice cultivars (Oryza sativa L.) in Malaysia. Water Air Soil Pollut 154:187–201

Ismail M, Suroto A, Abdullah S (2015) Response of Malaysian Local Rice Cultivars Induced by Elevated Ozone Stress. Environment Asia 8(1):86–93

Islam MT, Ashraf MA, Sattar MA (2007) Bio-monitoring study on tropospheric ozone using white clover at Bangladesh Agricultural University farming area. Progress Agric 18(2):215–222. ISSN 1017-8139

Kafiatullah AW, Ahmad SS, SRA S (2012) Ozone Biomonitoring in Pakistan using tobacco cultivar Bel-W3. Pak J Bot 44(2):717–723

Kharel K, Amgain LP (2010) Assessing the impact of ambient ozone on growth and yield of crop at Rampur, Chitwan. J Agri Environ 11:40–45

Klumpp A, Ansel W, Klumpp G, Vergne P, Sifakis N, Sanz MJ, Rasmussen S, Ro-Poulsen H, Ribas A, Penuelas J et al (2006) Ozone pollution and ozone biomonitoring inEuropean cities Part II. Ozone-induced plant injury and its relationship withdescriptors of ozone pollution. Atmos Environ 40(38):7437–7448

Köllner B, Krause GHM (2000) Changes in carbohydrate, leaf pigments and yield in potatoes induced by different ozone exposure regimes. Agric Ecosyst Environ 78:149–158

Kohut R (2005) Handbook for the assessment of foliar ozoneinjury on vegetation in the National Parks. http://www2.nature.nps.gov/air/permits/aris/networks/index.cfm

Kostka-Rick R (2002) Ozone biomonitoring in a local networkaround an automotive plant. In: Klumpp A, Fomin A, Klumpp G, Ansel W (eds) Bioindication and Air Quality in European Cities—Research, Application, Communication. Heimbach, Stuttgart, pp 243–248

Kumari S, Agrawal M, Tiwari S (2013) Impact of elevated CO2 and elevated O3 on Beta vulgaris L.: pigments, metabolites, antioxidants, growth and yield. Environ Pollut 174:279–288

Leisner CP, Ainsworth EA (2012) Quantifying the effects of ozoneon plant reproductive growth and development. Glob Chang Biol 18:606–616

Lesser VM, Rawlings JO, Spruill SE, Somerville MC (1990) Ozone effects on agricultural crops: statistical methodologies and estimated dose response relationships. Crop Sci 30:148–155

Li C, Meng J, Guo L, Jiang G (2016) Effects of ozone pollution on yield and quality of winter wheat under flixweed competition. Environ Exp Bot 129:77–84

Long S, Ainsworth E, Leakey A, Morgan P (2005) Global food insecurity. Treatment of major food crops with elevated carbon dioxide or ozone under large-scale fully open-air conditions suggests recent models may have overestimated future yields. Philos Trans R Soc B: Biol Sci 360:2011–2020

LRTAP (2004) Manual on methodologies and criteria for modelling and mapping critical loads and levels and air pollution effects, risks and trends. International Cooperative Programme on Mapping and Modelling under the UNECE Convention on Long-Range Transboundary Air Pollution. http://www.icpmapping.org

Matoušková L, Novotný R, Hůnová I, Buriánek V (2010) Visible foliar injury as a tool for the assessment of surface ozone impact on native vegetation: a case study from the Jizerské hory Mts. J For Sci 56(4):177–182

Mauzerall DL, Wang X (2001) PROTECTING AGRICULTURAL CROPS FROM THE EFFECTS OF TROPOSPHERIC OZONE EXPOSURE: Reconciling Science and Standard Setting in the United States, Europe, and Asia. Annu Rev Energy Environ 26:237–268

Manning WJ, Paoletti E, Sandermann H Jr, Ernst D (2011) Ethylenediurea (EDU): a research tool for assessment and verificationof the effects of ground level ozone on plants under natural condition. Environ Pollut 159:3283–3293

McGrath JM, Betzelberger AM, Wang S, Shook E, Zhu X-G, Long SP, Ainsworth EA (2015) An analysis of ozone damage to historical maize and soybean yields in the United States. PNAS 112(46):14390–14395

Meyer U, Kollner B, Willenbrink J, Krause GHM (2000) Effects of different ozone exposure regimes on photosynthesis, assimilates and thousand grain weight in spring wheat. Agric Ecosyst Environ 78(1):49–55

Michel A, Seidling W (eds) (2016) Forest Condition in Europe: 2016 Technical Report of ICP Forests. Report under the UNECE Convention on Long-Range Transboundary Air Pollution (CLRTAP). BFWDokumentation 23/2016. Vienna: BFW Austrian Research Centre for Forests, p 206

Middleton T (1956) Response of plants to air pollution. J Air Pollut Control Assoc 6(1):7–50

Mills G, Buse A, Gimeno B, Bermejo V, Holland M, Emberson L, Pleijel H (2007a) A synthesis of AOT40-based response functions and critical levels of ozone for agricultural and horticultural crops. Atmos Environ 41:2630–2643. https://doi.org/10.1016/j.atmosenv.2006.11.016

Mills G, Hayes F, Jones MLM, Cinderby S (2007b) Identifying ozone-sensitive communities of (semi-)natural vegetation suitable for mapping exceedance of critical levels. Environ Pollut 146:736–743. https://doi.org/10.1016/j.envpol.2006.04.005

Mills G, Hayes F, Simpson D, Emberson L, Norris D, Harmens H, Büker P (2011) Evidence of widespread effects of ozone on crops and (semi-)natural vegetationin Europe (1990-2006) in relation to AOT40- and flux-based risk maps. Glob Change Biol 17:592–613

Mills G, Harmens H, Wagg S, Sharps K, Hayes F, Fowler D, Sutton M, Davies B (2016) Ozone impacts on vegetation in a nitrogen enriched and changing climate. Environ Pollut 208 (898–908

Mina U, Kumar P, Varshney C (2010) Effect of ozone exposure on growth, yield and isoprene emission from tomato (Lycopersicon esculentum L.) plants. Veg Crops Res Bull 72:35–48

Mishra AK, Rai R, Agrawal SB (2013) Differential response of dwarf and tall tropical wheat cultivars to elevated ozone with and without carbon dioxide enrichment: growth, yield and grain quality. Field Crop Res 145:21–32

Morgan PB, Ainsworth EA, Long SP (2003) How does elevated ozone impact soybean? A meta-analysis of photosynthesis, growth and yield. Plant Cell Environ 26:1317–1328

Morgan PB, Mies TA, Bollero GA, Nelson RL, Long SP (2006) Season-longelevation of ozone concentration to projected 2050 levels under fully open-airconditions substantially decreases the growth and production of soybean. New Phytol 170:333–343

Nakicenovic N, Alcamo J, Davis G, de Vries B, Fenhann J, Gaffin S, Gregory K, Grubler A, Jung TY, Kram T, La Rovere EL, Michaelis L, Mori S, Morita T, Pepper W, Pitcher HM, Price L, Riahi K, Roehrl A, Rogner H-H, Sankovski A, Schlesinger M, Shukla P, Smith SJ, Swart R, van Rooijen S, Victor N, Dadi Z (2000) Emissions Scenarios: a Special Report of Working Group III of the Intergovernmental Panel on Climate Change. Cambridge Univ. Press, New York, p 599

Ollerenshaw JH, Lyons T (1999) Impacts of ozone on growth field grown winter wheat. Environ Pollut 106:67–72

Ollerenshaw JH, Lyons T, Barnes JD (1999) Impacts of ozone on the growth and yield of field-grown winter oilseed rape. Environ Pollut 104:171–179

Osborne SA, Mills G, Hayes F, Ainsworth EA, Buker P, Embersen L (2016) Has the sensitivity of soybean cultivars to ozone pollution increased with time? An analysis of published dose–response data. Glob Chang Biol 22:3097–3111

Pandey AK, Majumder B, Keski-Saari S, Kontunen-Soppela S, Pandey V, Oksanen E (2014) Differences in responses of two mustardgenotypes to ethylenediurea (EDU) at high ambient ozone concentrations in India. Agric Ecosyst Environ 196:158–166

Pang J, Kobayashi K, Zhu J (2009) Yield and photosynthetic characteristics of flag leaves in Chinese rice (Oryza sativa L.) varieties subjected to free-air release of ozone. Agric Ecosyst Environ 132:203–211

Pasqualini S, Paoletti E, Cruciani G, Pellegrino R, Ederli L (2016) Effects of different routes of application on ethylenediurea persistence in tobacco leaves. Environ Pollut 212:559–564

Paoletti E, Manning WJ, Spaziani F, Tagliaferro F (2007a) Gravitational infusion of ethylenediurea into trunks protected adult European ash tree (*Fraxinus excelsior* L.) from foliar ozone injury. Environ Pollut 145:869–873

Paoletti E, Contran N, Manning WJ, Tagliaferro F (2007b) Ethylenediurea affects the growth of ozone sensitive and tolerantash (Fraxinus excelsior) trees under ambient ozone conditions. Sci World J 7(S1):128–133

Paoletti E, Contran N, Manning WJ, Castagna A, Ranieri A, Tagliaferro F (2008) Protection of ash (Fraxinus excelsior L.) trees from ozoneinjury by ethylenediurea (EDU): roles of biochemical changes anddecreased stomatal conductance in enhancement of growth. Environ Pollut 155:464–472

Paoletti E, Contran N, Manning WJ, Ferrara AM (2009) Use of the antiozonant ethylenediurea (EDU) in Italy: verification of the effectsof ambient ozone on crop plants and trees and investigation of EDU's mode of action. Environ Pollut 157:1453–1460

Paoletti E, Castagna A, Ederli L, Pasqualini S, Ranieri A, Manning WJ (2014) Gene expression in snapbeans exposed to O3 and protectedby ethylenediurea. Environ Pollut 193:1–5

Persson K, Danielsson H, Sellden G et al (2003) The effects of tropospheric ozone and elevated carbon dioxide on potato (Solanum tuberosum L. cv. Bintje) growth and yield. In: Detecting Environmental Change – Science and Society Conference. Elsevier Science Bv, London, pp 191–201

Piikki K, Sellden G, Pleijel H (2004) The impact of tropospheric O3 on leaf number duration and tuber yield of the potato (Solanum tuberosum. L.) cultivars Bintje and Kardal. Agric Ecosyst Environ 104:483–492

Pleijel H, Norberg PA, Sellden G, Skarby L (1999) Tropospheric ozone decreases biomass production in radish plants (Raphnus sativus) grown in rural South West Sweden. Environ Pollut 106:143–147

Pleijel H, Skärby L, Wallin G, Selldén G (1995) A process oriented explanation of the non-linear relationship between grain yield of wheat and ozone exposure. New Phytol 131:241–246

Pleijel H, Danielsson H, Gelang J, Sild E, Sellde'n G. (1998) Growth stage dependence of the grain yield response to ozone in springwheat (*Triticum aestivum* L.). Agriculture. Ecosyst Environ 70:61–68

Pleijel H, Danielsson H, Ojanperä K, Temmerman LD, Högy P, Badiani M, Karlsson PE (2004) Relationships between ozone exposure and yield loss in European wheatand potato-a comparison of concentration- and flux-based exposure indices. Atmos Environ 38:2259–2269

Pleijel H, Danielsson H, Simpson D, Mills G (2014) Have ozone effects on carbon sequestration been overestimated? A new biomass response function for wheat. Biogeosciences 11:4521–4528. https://doi.org/10.5194/bg-11-4521-2014

Rai R, Agrawal M, Agrawal SB (2007) Assessment of yield losses in tropical wheat using open top chambers. Atmos Environ 41:9543–9554

Reid CD, Fiscus EL (2008) Ozone and density affect the response of biomass and seed yield to elevated CO2 in rice. Glob Chang Biol 14:60–76

Rai R, Agrawal M (2008) Evaluation of physiological and biochemical responses of two rice (Oryza sativa L.) cultivars to ambient air pollution using open top chambers at rural site in India. Sci Total Environ 407:679–691

Rai R, Agrawal M (2014) Assessment of competitive ability of two Indian wheat cultivars under ambient O3 at different developmental stages. Environ Sci Pollut Res 21:1039–1053

Rai R, Agrawal M, Choudhary KK, Agrawal S, Emberson L, Büker P (2015) Application of ethylene diurea (EDU) inassessing the response of a tropical soybean cultivar to ambient O3: Nitrogen metabolism, antioxidants, reproductive development and yield. Ecotoxicol Environ Saf 112:29–38

Reidmiller DR et al (2009) The influence of foreign vs. North Americanemissions on surface ozone in the U.S. Atmos Chem Phys 9:5027–5042

Ribas A, Peñuelas J (2002) Ozone bioindication in Barcelonaand surrounding area of Catalonia. In: Klumpp A, Fomin A, Klumpp G, Ansel W (eds) Bioindication and Air Quality in European Cities—Research, Application, Communication. Heimbach, Stuttgart, pp 221–225

Ribas A, Peñuelas J (2003) Biomonitoring of tropospheric ozone phytotoxicity in rural Catalonia. Atmos Environ 37:63–71

Richards BL, Middleton JT, Hewitt WB (1958) Air pollution with reference to agronomic crops. Agron J 50:559–561

Sanz MJ, Calatayud V (2011) Ozone Injury in European Forest Species. http://ozoneinjury.org

Sanz MJ, Calatayud V (2013) Ozone injury in European forest species. Accessed 04.12.13. http://www.ozoneinjury.org/

Sanz J, Gonzales I, Calvete-Sogo H, Bermejo V (2014) Ozone and nitrogen effects on yield and nutritive quality of the annual legume Trifolium cherleri. Atmos Environ 94:765–772

Sarkar A, Agrawal SB (2010) Elevated ozone and two modern wheat cultivars: an assessment of dose dependent sensitivity with respect to growth, reproductive and yield parameter. Environ Exp Bot 69:328–337

Sarkar A, Singh AA, Agrawal SB, Ahmed A, Rai SP (2015) Cultivar specific variations in antioxidative defense system, genome and proteome of two tropical rice cultivars against ambient and elevated ozone. Ecotoxiocol Environ Saf 115:101–111

Sarkar A, Datta S, Singh P (2017) Tropospheric ozone pollution, agriculture, and food security. In: Singh RP, Singh A, Srivastava V (eds) Environmental issues surrounding human overpopulation. IGI Global, Hershey, pp 234–252

Schenone G, Lorenzini G (1992) Effects of regional air pollution on crops in Italy. Agric Ecosyst Environ 38:51–59

Schaub M, Calatayud V (2013) Assessment of visible foliar injury induced by ozone. In: Ferretti M, Fisher R (eds) Forest monitoring, methods for terrestrial investigations in Europe with an overview of North America and Asia, developments in environmental science 12. Elsevier, Oxford, pp 205–221

Schaub M, Calatayud V, Ferreti M, Brunialti G, Lⲥovblad G, Krause G, Sanz MJ (2010) Assessment of ozone injury. Manual on methods and criteria for harmonized sampling, assessment, monitoring and analysis of the effects of air pollution on forests. UNECE ICP Forests Programme Co-ordinatingCentre, Hamburg, p 1e22. Available at: http://icp-forests.net/

Scheepers CCW, Strasser RJ, Krüger GHJ (2013) Effect of Ozone on Photosynthesis and Seed Yield of Sensitive (S156) and Resistant (R123) Phaseolus Vulgaris L. enotypes in Open-Top

Chambers. In: Photosynthesis Research for Food, Fuel and the Future. Advanced Topics in Science and Technology in China. Springer, Berlin, Heidelberg

Seiler LS (2012) Effectiveness of *Ailanthus altissima* as a Bioindicator of OzonePollution (M.S. thesis). The Pennsylvania State University

Seinfeld JH, Pandis SN (2012) Atmospheric chemistry and physics: from air pollution to climate change. John Wiley and Sons, New York

Shi G, Yang L, Wang Y, Kobayashi K, Zhu J, Tang H, Pan S, Chen T, Liu G, Wang Y (2009) Impact of elevated ozone concentration on yield of four Chinese rice cultivars under fully open-air field conditions. Agric Ecosyst Environ 131:178–184

Simpson D, Emberson L, Ashmore M, Tuovinen J-P (2007) A comparison of two different approaches for mapping potential ozone damage to vegetation. A model study. Environ Pollut 146:715–725

Singh AA, Singh S, Agrawal M, Agrawal SB (2015) Assessment of Ethylene Diurea-Induced Protection in Plants Against OzonePhytotoxicity. Rev Environ Contam Toxicol 233:129–184

Singh S, Agrawal SB (2009) Use of ethylenediurea (EDU) in assessing the impact of ozone on growth and productivity of five cultivars of Indian wheat (Triticum aestivum L.) Environ Monit Assess 159:125–141

Singh S, Agrawal SB (2011a) Cultivar specific response of soybean (Glycine max L.) to ambient and elevated concentrations of ozone under open top chambers. Water Air Soil Pollut 217:283–302

Singh S, Agrawal SB (2011b) Ambient ozone and two black gram cultivars: an assessment of amelioration by use of ethylenediurea. Acta Physiol Plant 33(6):2399–2411

Singh P, Agrawal M, Agrawal SB (2009) Evaluation of physiological growth and yield responses of a tropical oil crop (Brassica campestris L. var. Kranti) under ambient ozone pollution at varying NPK levels. Environ Pollut 157:871–880

Singh E, Tiwari S, Agrawal M (2010) Variability in antioxidant and metabolite levels, growth and yield of two soybean varieties: an assessment of anticipated yield losses under projected elevation of ozone. Agric Ecosyst Environ 135:168–177

Singh AA, Agrawal SB, Shahi JP, Agrawal M (2014) Assessment of growth and yield losses in two Zea mays L. cultivars (quality protein maize and nonquality protein maize) under projected levels of ozone. Environ Sci Pollut Res 21:2628–2641

Sinha B, Singh SK, Maurya Y, Kumar V, Sarkar C, Chandra BP, Sinha V (2015) Assessmentof crop yield losses in Punjab and Haryana using 2 years of continuous insitu ozone measurements. Atmos Chem Phys 15:9555–9576

Sitch S, Cox PM, Collins WJ, Huntingford C (2007) Indirect forcing of climate change through ozone effects on the land carbon sink. Nature 448:791–793

Smith G (2012) Ambient ozone injury to forest plants in Northeast and NorthCentral USA: 16 years of biomonitoring. Environ Monit Assess 184:4049–4065

Smith GC, Coulston JW, O'Connell BM (2008) Ozone bioindicators and forest health: A guide to the evaluation, analysis, and interpretation of ozone injury data in the Forest Inventory and Analysis Program. Gen. Tech. Rep.NRS-34. Newtown Square, PA: US Department of Agriculture. Forest Service, Northern Research Station, p 34

Tang H, Takigawa M, Liu G, Zhu J, Kobayashi k (2013) A projection ofozone-induced wheat production loss in China and India for theyears 2000 and 2020 with exposure-based and flux-based approaches.Glob. Chang Biol 19(9):2739–2752

Taia W, Basahi J, Hassan I (2013) Impact of ambient air on physiology, pollen tube growth, pollen germination and yield of pepper (Capsicum annum L.) Pak J Bot 45(3):921–923

Thanacharoenchanaphas K, Rugchati O (2010) Adverse effects of elevated ambient ozone on yield and protein loss of three Thai soybean cultivars. Int J Environ Rural Dev 1–2:12–17

Tiwari S (2017) Ethylenediurea as a potential tool in evaluating ozonephytotoxicity: a review study on physiological, biochemical and morphological responses of plants. Environ Sci Pollut Res 24:14019–14039. https://doi.org/10.1007/s11356-017-8859-y

Tiwari S, Agrawal M (2009) Protection of palak (Beta vulgaris L. var Allgreen) plants from ozone injury by ethylenediurea (EDU): Roles of biochemical and physiological variations in alleviating the adverse impacts. Chemosphere 75:1492–1499

Tiwari S, Agrawal M (2010) Effectiveness of different EDU concentrations in ameliorating ozone stress in carrot plants. Ecotoxicol Environ Saf 73(5):1018–1027

Tiwari S, Agrawal M (2011) Evaluation of effect of ambient air pollutants on morphological characteristics and nutrient contents of radish plants grown in open top chambers. Environmental pollution: a threat to living world. Jaspal Prakashan, Patna, pp 32–36

Tiwari S, Agrawal M, Manning WJ (2005) Assessing the impacts of ambient ozone on growth and productivity of two cultivars of wheat in India using three rates of applications of ethylenediurea (EDU). Environ Pollut 138:153–160

Tiwari S, Agrawal M, Marshall F (2006) Evaluation of ambient airpollution impact on carrot plants at a suburban site using open topchamber. Environ Monit Assess 266:15–30

Tiwari S, Agrawal M, Marshal FM (2010) Seasonal variations in adaptational strategies of Beta vulgaris L. plants in response to ambient air pollution: Biomass allocation, yield and nutritional quality. Trop Ecol 51(2S):353–363

Thwe AA, Vercambre G, Gautier H, Gay F, Phattaralerphong J, Kasemsap P (2014) Response of photosynthesis and chlorophyll fluorescence to acute ozone stress in tomato (Solanumlycopersicum Mill.) Photosynthetica 52:105–116

Tomer R, Bhatia A, Kumar V, Kumar A, Singh R, Singh B, Singh SD (2015) Impact of elevated ozone on growth, yield and nutritional quality of two wheat species in Northern India. Aerosol Air Qual Res 15:329–340

Tong D, Mathur R, Schere K, Kang D, Yu S (2007) The use of air quality forecasts to assess impacts of air pollution on crops: methodology and case study. Atmos Environ 41:8772–8784

Tonneijk AEG, van Dijk CJ (1997) Assessing effects of ambient ozone on injury and growth of Trifolium subterraneum at four rural sites in Netherlands with ethylenediure (EDU). Agric Ecosyst Environ 65:79–88

Tripathi R, Agrawal SB (2012) Effects of ambient and elevated level of ozone on Brassica campestris L. with special reference to yield and oil quality parameters. Ecotoxicol Environ Saf 85:1–12

Tripathi R, Agrawal SB (2013) Interactive effect of supplemental ultraviolet B and elevated ozone on seed yield and oil quality of two cultivars of linseed (Linum usitatissimum L.) carried out in open top chambers. J Sci Food Agric 93:1016–1025

UNECE (1999) Air pollution and vegetation. Annual Report 1998/1999. ICP-Vegetation Coordination Centre, CEH Bangor, University of Wales, Bangor

Unsworth MH, Heagle AS, Heck WW (1984) Gas exchange in open-top field chambers: measurement and analysis of atmospheric resistances to gas exchange. Atmos Environ 18:373–380

Volk M, Obrist D, Novak K, Giger R, Bassin S, Fuhrer J (2011) Subalpine grassland carbon dioxide fluxes indicate substantial carbon losses under increased nitrogen deposition, but not at elevated ozone concentration. Glob Chang Biol 17:366–376

Van Dingenen R (2009) The global impact of O3 on agricultural cropyields under current and future air quality legislation. Atmos Environ 43:604–618

Wahid A, Milne E, Shamsi SRA, Ashmore MR, Marshall FM (2001) Effects of oxidants on soybean growth and yield in Pakistan Punjab. Environ Pollut 113:271–280

Wahid A (2006) Influence of atmospheric pollutants on agriculture in developing countries: a case study with three new wheat varieties in Pakistan. Sci Total Environ 371:304–313

Wahid A, Maggs R, Shamasi SRM, Bell JNB, Ashmore MR (1995) Effects of air pollution on rice yield in the Pakistan Punjab. Environ Pollut 90:323–329

Wang X, Zheng Q, Yao F, Chen Z, Feng Z, Manning WJ (2007) Assessing the impact of ambient ozone on growth and yield of rice (Oryza sativa L.) and wheat (Triticum aestivum L.) cultivar grown in Yangtze Delta, China, using three rates of application of Ethylenediurea (EDU). Environ Pollut 148:390–395

WHO (2000) Air quality guidelines for Europe. World Health Organization Regional Office for Europe. Copenhagen, Denmark

Wan H, Zhang X, Zwiers F, Emori S, Shiogama H (2013) Effect of data coverage on the estimation of mean and variability of precipitation at global and regional scales. J Geophys Res **118**:534–546

Wan W et al (2014) Ozone and ozone injury on plants in andaround Beijing, China. Environ Pollut 191:215–222

Wang X, Mauzerall DL (2004) Characterizing distributions of surfaceozone and its impact on grain production in China, Japan and SouthKorea: 1990 and 2020. Atmos Environ 38:4383–4402

Wang X, Zhang Q, Zheng F, Zheng Q, Yao F, Chen Z, Zhang W, Hou P, Feng Z, Song W, Feng Z, Lu F (2012) Effects of elevated O3 concentration on winter wheat and rice yields in the Yangtze River Delta, China. Environ Pollut 171:118–125

Wat EKW (1978) Prevention of ozone injury to plants by a new protectant chemical. Phytopathology 68:1225–1229

Weidensaul TC (1980) N-[2-(2-oxo-1-imidazolinidyl)ethyl]-Nphenylurea as a protectant against ozone injury to laboratory fumigated pinto bean plants. Phytopathology 70:42–45

Wilkinson S, Gina Mills G, Rosemary Illidge R, Davies WJ (2011) How is ozone pollution reducing our food supply? J Exp Bot Adv Access 20. https://doi.org/10.1093/jxb/err317

Xin Y, Yuan X, Shang B, ManningWJ YA, Wang Y, Feng Z (2016) Moderate drought did not affect the effectiveness of ethylenediurea(EDU) in protecting *Populus cathayana* from ambient ozone. Sci Total Environ 569–570:1536–1544

Yamaguchi M, Hoshino H, Inada H, Akhtar N, Sumioka C, Takeda K, Izuta T (2014) Evaluation of the effects of ozone on yield of Japanese rice (*Oryza sativa* L.) based on stomatal ozone uptake. Environ Pollut 184:472–480

Yi F, Jiang F, Zhong F, Zhou X, Ding A (2016) The impacts of surface ozone pollution on winter wheat productivity in China- an econometric approach. Environ Pollut 208:326–335

Yonekura T, Izuta T (2017) Effects of Ozone on Japanese Agricultural Crops. In: Izuta T (ed) Air pollution impacts on plants in East Asia. Springer, Tokyo, pp 57–71

Yuan X, Calatayud V, Jiang L, Manning WJ, Hayes F, Tian Y, Feng Z (2015) Assessing the effects of ambient ozone in China on snap bean genotypes by using ethylenediurea (EDU). Environ Pollut 205:199–208

Zhao Y, Bell JNB, Wahid A, Power SA (2011) Inter andintra-specific differences in the response of Chinese leafyvegetables to ozone. Water Air Soil Pollut 216:451–462

Zhang W, Wang G, Liu X, Feng Z (2014) Effects of elevated O3 exposure on seed yield, N concentration and photosynthesis of nine soybean cultivars (Glycine max (L.) Merr.) in Northeast China. Plant Sci 226:172–181

Zhu X, Feng Z, Sun T, Liu X, Tang H, Zhu J, Guo W, Kobayashi K (2011) Effects of elevated ozone concentration on yield of four Chinese cultivars of winter wheat under fully open-air field conditions. Glob Chang Biol 17:2697–2706

Zouzoulas D, Spyridon DK, Vassiliou G, Vardavakis E (2009) Effects of ozone fumigation on cotton (Gossypium hirsutum L.) morphology, anatomy, physiology, yield and qualitative characteristics of fibers. Environ Exp Bot 67:293–303

Chapter 5
Mitigation of Ozone Stress

Abstract The photochemical reactions leading to O_3 formation and the variables on which these reactions depend are undergoing rapid alterations owing to the present climate change scenario. The multifarious set-up related to O_3 formation in the troposphere makes it difficult to check the continuously increasing concentration of O_3 around the globe. O_3 concentration has already crossed the standard limit for vegetation set by European Union (EU) in most of the parts of the World, which is evident by a number of O_3 induced crop yield reduction studies. Therefore the demand of the time is to develop certain strategies that will help in alleviating the deleterious effects of O_3 on plant performance. This target can be achieved by adopting different approaches such as improved agronomic practices; selection of O_3 resistant cultivars, improving photosynthetic efficiencies of O_3 exposed plants etc. Several strategies have been followed to achieve these targets, important ones being CO_2 fertilization and soil nutrient amendments. In addition to this, air quality management practices using CH_4 emission control is also considered to be an important strategy in minimizing O_3 induced stress in plants. It has been shown that CO_2 fertilization increased the carbon input which can be incorporated in plant biomass and subsequently helps in maintaining yield of plants exposed to O_3 stress. Treatment of additional nutrient helps in the repair of the O_3 injured plants thus sustaining their yield. As apparent through a number of studies, these strategies have proved quite efficient in partially mitigating O_3 stress in plants. However, more experimentation is required before confirming the use of these approaches in mitigating O_3 injury and implementing them in daily agricultural practices.

Keywords Mitigation · CO_2 fertilization · Nutrient amendments · Alleviation

Contents

© Springer International Publishing AG 2018
S. Tiwari, M. Agrawal, *Tropospheric Ozone and its Impacts on Crop Plants*,
https://doi.org/10.1007/978-3-319-71873-6_5

1   Introduction

The global threat to food security due to O_3 pollution and its interaction with climate change has been well documented (Tai et al. 2014; Tian et al. 2016; Sarkar et al. 2017; Wilkinson et al. 2012). With the present climate change scenario providing more favourable conditions for O_3 formation, average O_3 concentration as high as 50 ppb has been recorded from several parts of the Northern hemisphere during the growing seasons of important crops (Tang et al. 2013; Feng et al. 2015). Significant reductions recorded in important food crops like wheat, rice, maize and soybean clearly indicate that global food security is threatened by the current O_3 concentration and this threat is expected to intensify in the future (Avnery et al. 2011a, b; Osborne et al. 2016; Danh et al. 2016). With the current trends of population growth, the World food production will have to be doubled to feed the total population by 2025 (Tiwari et al. 2017). IPCC fifth assessment report (2013) indicated that the developing Asian and African countries are most likely to face the food security threat not only because of their unchecked population growth rate, but also due to unplanned urbanization and industrialization which provide favourable conditions for O_3 production. The individual experimental studies conducted in these regions recorded significant yield reductions in important staple crops (as discussed in the previous chapter) providing sufficient evidences to justify the global food security threat.

O_3 is a multifarious molecule whose concentration in the atmosphere is difficult to check due to multiple sources leading to its formation in the troposphere (Fig. 5.1). In addition to the fact that current climate change scenario creates conditions favourable for the increased rate of emission of different O_3 precursors, anthropogenic activities also add to the already intensified problem of the increasing concentration of ground level O_3 (Fig. 5.1). Since O_3 is central to the chemistry of troposphere

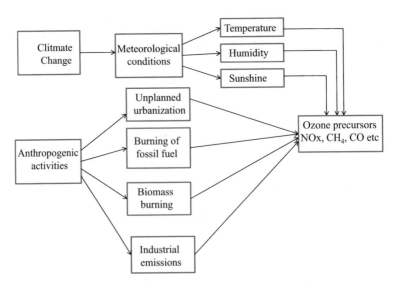

Fig. 5.1 Factors affecting the emission of ozone precursors

owing to it role in the initiation of photochemical oxidation processes, the climate induced meteorological conditions also play an important role in O_3 formation.

2 O₃ Mitigation Strategies

The variable sources of O_3 formation clearly suggest that large amount of efforts are required to check the increasing O_3 concentration and keep it under control in the troposphere (Monks et al. 2015). It has been observed that O_3 reductions via mitigation of conventional pollutant precursors (NOx, CO, NMVOCs) would prevent significant additional future yield reductions (Van Dingenen et al. 2009; Avnery et al. 2011b). However, even with strict emission controls, global year 2030 losses are predicted to be sufficiently high, especially for the O_3 sensitive crops (Avnery et al. 2011b). Since the productivity losses for crop plants due to increasing O_3 concentrations are difficult to ignore, it is essential to develop certain strategies that can mitigate O_3 injury in plants to some extent. In addition to air quality management through targeting the short lived O_3 precursors, it is essential to explore supplemental strategies to reduce O_3 induced crop losses (Avnery et al. 2013). O_3 mitigation studies therefore have adopted different strategies which can be studied under two roughly categorized sections (Fig. 5.2). The first strategy is to improve the agricultural production by reducing the associated yield reductions caused by the exposure of the crops to surface O_3. This strategy involves two approaches, the first one targets towards the gradual reduction in the emission rates of important O_3 precursors such as methane (CH_4) (Avnery et al. 2013), while the other focuses on compensating the induced yield reductions through selected agronomic practices such as CO_2 fertilization (Kumari et al. 2013) and soil nutrient amendments (Singh et al. 2015).

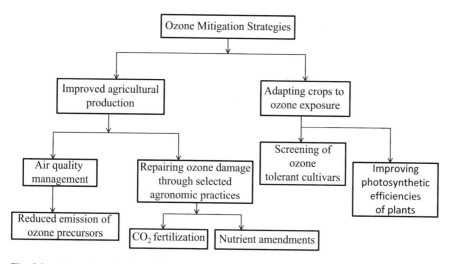

Fig. 5.2 A flow-chart showing different ozone mitigation strategies

The second strategy of O_3 mitigation focuses on adapting crops to O_3 exposure by selecting cultivars with demonstrated O_3 resistance (Avnery et al. 2013) (Fig. 5.2).

2.1   Improving Agricultural Production

The most important challenge that is to be faced in near future is the increasing global grain demand that is projected to be 50% more than the current value by 2030, if no further environmental degradation is taken into consideration (Alexandratos and Bruinsma 2012). Although various agronomic and breeding efforts in the past decades have been implemented with positive results, a further ability to increase or even sustain crop yield or quality is required in view of the rapid global environmental change (Tester and Langridge 2010). Since O_3 has been well proved to be an abiotic stress responsible for causing significant losses in agricultural production, O_3 mitigation strategies should be adopted to reduce the productivity losses. The coming sections of the chapter deal with the important strategies which can be useful in modifying the plant response resulting in lesser productivity losses.

2.1.1   Air Quality Management

As discussed in the earlier chapter, O_3 is produced in the troposphere by photochemical reactions between NOx, CO, CH_4 and NMVOCs emitted in the troposphere via different sources (Bortolin et al. 2014). Studies done for estimating the O_3 induced damage to agricultural production has paid less attention to the role of air quality management, which aims at reducing the emission rates of O_3 precursors, thus placing a check on the process of O_3 formation in the troposphere (Ghude et al. 2016). One of the major requirements for the air quality management for a particular region is to identify the relative importance of the emitted precursors. For instance, NOx is considered to play a pivotal role in O_3 formation followed by CH_4 and VOCs (Ghude et al. 2016; Bortolin et al. 2014; Avnery et al. 2013).

Ghude et al. (2016) analysed the role of O_3 precursors, mainly NOx and VOCs in mitigating O_3 injury in plants. Using a regional chemistry transport model WRF-CHEM (version 3.2.2) coupled with AOT40 exposure indices, risks imposed upon agricultural production of two major crops in India (wheat and rice) were estimated. This study by Ghude et al. (2016) used three approaches; baseline simulations with present day anthropogenic NOx and VOCs emissions, without anthropogenic NOx and VOCs emissions and a comparison of the two simulations to assess the response of NOx and VOCs mitigation action on O_3 induced wheat and rice production losses in India. The comparative results showed that simulations with no anthropogenic NOx emissions led to a 93% reduction in O_3 induced crop yield losses in case of rice and 87% in case of wheat as compared to the baseline scenario. It was observed that in case of no anthropogenic VOCs emissions only a small percentage of O_3 induced yield losses were compensated. The response of VOC mitigation scenario resulted in about 22% reduction in rice losses and 26% in case of wheat as compared to the baseline scenario (Ghude et al. 2016). A comparison in the mitigation of yield losses between no anthropogenic NOx and VOC emission simulations showed that absence of NOx could mitigate the O_3 induced injury to a larger extent as compared to the absence of VOCs. This suggests that NOx has a more important role to play in O_3 formation in troposphere as compared to VOCs (Ghude et al. 2016).

Based upon the similar concept of mitigation of O_3 induced agricultural losses by regulating air quality, Avnery et al. (2013) evaluated the increase in global agricultural production via CH_4 emission control. Besides acting as an important O_3 precursor, CH_4 also acts as an important greenhouse gas, second after CO_2 (Forster et al. 2007). Anthropogenic CH_4 from agricultural and industrial sectors contribute ~ 0.7 W^{-2} to climate forcing and 4 ppbv to surface O_3 in the year 2030 CLE (current legislation emission) (Fiore et al. 2008). Therefore, CH_4 abatement provides a double profit opportunity for both the climate change and air pollution mitigation goals, as CH_4 emission reduction would help reduce the radiative forcing of climate along with achieving agricultural benefits associated with surface O_3 reductions (Shindell et al. 2012).

Avnery et al. (2013) used multi decadal full chemistry transient simulations of the MOZRAT-2 global CTM (Horowitz et al. 2003) to quantify the crop production improvements possible by applying the strategy of controlling CH_4 emissions relative to the CLE scenario. Under CLE, global anthropogenic emissions are projected to increase by 35% between 2000 and 2030, assuming full implementation of the legislation controlling the emissions of traditional air pollutants (Cofala et al. 2007). The above mentioned model was used to evaluate the response of surface O_3 to future CH_4 emission from 2000 to 2030 under CLE scenario and reduced CH_4 scenario (CH_4-red) as per scenario B of IPCC (2013). The CLE and CH_4-red are not in a steady state, therefore the full benefits of gradually increasing CH_4 reductions will not be realized by 2030 due to relatively long lifetime of CH_4 (~12 years). Using MOZRAT- 2 model, it was predicted that in CLE scenario, global anthropogenic emissions of CH_4 will change by 29% from 2005 to 2030 (Cofala et al. 2007; Fiore et al. 2008). CH_4 – red scenario predicted a reduction of 30% in global anthropogenic CH_4 emission relative to the CLE baseline (Avnery et al. 2013). It has been shown that the reduction in CH_4 emission as per the CH_4- red scenario decreased the global year O_3 exposure by 9.1–11.9% during the wheat growing season, especially in the regions where O_3 exposure to CH_4 control is greatest (particularly in South and East Asia).

The strong positive response of O_3 mitigation via checking CH_4 emissions during the growing season of wheat in South Asia dominates the global crop production improvements (Avnery et al. 2013). The same scenario predicts reductions of 8.9–10.6% and 7.9–9.7% in the O_3 exposure during maize and soybean seasons, respectively (Avnery et al. 2013). Based upon these observations, the crop production losses (CPL) under CLE and CH_4-red scenarios were dominated by wheat, when calculated under both AOT40 and W126 exposure indices, accounting for 77–85% of global losses of all the three crops. Checking of the anthropogenic CH_4 emission examined here would lead to significant reduction in CPL with over 85% of crop loss improvements due to wheat yield scenarios (Avnery et al. 2013). Avnery et al. (2013) predicted that relative to the values of the year 2000, wheat yield showed an improvement of 3.7–19%, whereas in case of maize and soybean, the improvement was only 1%. The CH_4 emission control in CH_4- red scenario could increase the combined annual production of maize, soybean and wheat by 2.0–8.3% in 2030 as compared to the annual production under CLE scenario. These values represent a prevention of 10–45% of O_3 induced CPL that are otherwise projected to occur in 2030 under CLE scenario (Avnery et al. 2013).

2.1.2   Repairing the O_3 Induced Damage in Plants by Using Different Agronomic Practices

(a) CO_2 Fertilization

Elevated atmospheric CO_2 is one of the most important components of climate change (IPCC 2013). The global atmospheric CO_2 concentration has shown an increase of 43% from the pre- industrial level of 280 ppm in 1750 to a present level of 400 ppm, with an increase of 1.55 ppm CO_2 per year (globally) over the past 55 years (Xu et al. 2015). As a result of uncontrolled combustion of fossil fuels, deforestation and change in land use pattern, atmospheric CO_2 is predicted to reach a level of 730–1050 ppm by 2100 (IPCC 2013). The increasing CO_2 concentration in the atmosphere is considered to be beneficial as it positively affects the plant's performance and results in increased plant productivity (Degener 2015; Broberg 2015; Hogy et al. 2013; Weigel and Manderscheid 2012). Elevated CO_2 can stimulate plant growth by providing additional carbon, the phenomenon termed as CO_2 fertilization effect. The CO_2 fertilization assists the plants by mitigating wide-ranging abiotic stresses including O_3 through certain mechanisms which are still under investigation (AbdElgawad et al. 2016). Researchers have shown that elevated CO_2 brings about a few modifications in targeted plant's physiological and biochemical processes which help in partial amelioration of O_3 injury (Hager et al. 2016; Xu et al. 2015). Plants growing under stress generally accumulate unnecessary ROS, which is in excess of the scavenging capacity of the plant's intrinsic defense system (enzymatic and non enzymatic antioxidants) (Sharma et al. 2012; Tripathy and Oelmüller 2012; Gill and Tuteja 2010). It has been suggested that elevated CO_2 helps in detoxifying ROS produced under O_3 stress (Kumari et al. 2013).

Elevated CO_2 not only affects the crucial biological processes such as photosynthesis, respiration, antioxidant system and other secondary metabolic activities in plants (Peñuelas et al. 2013; Singh and Agrawal 2015), but also influences the gene expression of the biological processes that may be modified by elevated CO_2 (Zinta et al. 2014; Jagadish et al. 2014; Peñuelas et al. 2013). Zinta et al. (2014) showed that in *Arabidopsis thaliana* L. Columbia, the down regulation of PS II genes in plants exposed to stress conditions was partially mitigated under elevated CO_2

(730 ppm). Similarly elevated CO_2 induced a decline in MDA content of the stress exposed plants of *A. thaliana* L. Columbia suggesting reduced oxidative membrane damage. Kotchoni et al. (2006) demonstrated that overexpression of aldehyde dehydrogenase genes (ALDHs) provided protection against lipid peroxidation and reduced MDA content by aldehyde detoxification. Transcriptome analysis done by Zinta et al. (2014) revealed that ALDHs levels were highly up regulated in *A. thaliana* L. Columbia when grown at elevated CO_2 as compared to plants grown under stress conditions. Moreover, MDA itself acts as a modifying reagent and can affect gene expression and activate defense responses (Weber et al. 2004).

The stimulation of light saturated photosynthetic CO_2 assimilation rate is a general response of plants to CO_2 enrichment with an average 31% increase observed by Ainsworth and Rogers (2007). However, the magnitude of response varies with different plant functional types (PFTs), with maximum for trees and C_3 grasses, moderate for shrubs, C_3 and C_4 crops and legumes and minimum for C_4 grasses (Xu et al. 2015; Ainsworth and Rogers 2007). In addition to this, rate of photosynthesis in plants under elevated CO_2 also depend upon the environmental conditions like nutrient and water resource availability (Xu et al. 2015). Markelz et al. (2014) demonstrated that under abundant nitrogen supply, elevated CO_2 increased the photosynthetic rate by 82% in *A. thaliana* leaves. However, no significant effect on the midday net photosynthetic rate was observed in soybean under elevated CO_2 conditions (Bunce 2014). This unusual observation was explained by the fact that at high photosynthetic photon flux density (PPFD), the net photosynthetic rate was delimited by the low carboxylation capacity of RuBisCO (Bunce 2014). Elevated CO_2 may increase the levels of antioxidants including ascorbate, phenols and alkaloids and the activities of some antioxidant enzymes such as CAT and SOD leading to a decline in ROS level (Mishra and Agrawal 2014; Zinta et al. 2014). Kumari et al. (2013) observed increased levels of ascorbate and phenol in *Beta vulgaris* L. var. Allgreen exposed to elevated CO_2 (570 ppm), while increments in ascorbate (ASC), glutathione (GSH) and ASC/GSH were recorded in *Lolium perenne* and *Medicago lupulina* at elevated CO_2 concentration (Farfan-Vignolo and Asard 2012). Reduction in oxidative stress was recorded in *Zingiber officinale* (Ghasemzadeh et al. 2010), *Catharanthus roseus* (Singh and Agrawal 2015), *Vigna radiate* (Mishra and Agrawal 2014) and *A. thaliana* (Zinta et al. 2014) under CO_2 fertilization. Stimulation of antioxidant machinery may also play an important role in dealing with the biological process of senescence (Hodges and Forney 2000). Under elevated CO_2, there is an increase in total non structural carbohydrates including starch and sugars (glucose, fructose and sucrose) while the structural carbohydrates cellulose), lignin and lipids remain unaffected (Markelz et al. 2014). Nitrogen assimilation, however, may be enhanced by elevated CO_2 (Ribeiro et al. 2012). Guo et al. (2013) reported that elevated CO_2 may promote the activities of nitrogen assimilation and transamination related enzymes such as glutamate oxoglutarate aminotransferase (GOGAT) and glutamate oxalate transaminase (GOT) in *Medicago truncatula*.

Due to the stimulatory effect of elevated CO_2 on plant growth and productivity, CO_2 fertilization can be used for modifying the negative effects of O_3 stress on plants (Kumar 2016; Kumari et al. 2013). Theoretically, an increase in the atmospheric CO_2 upto 550 and 700 ppm has the potential to increase the productivity of C_3 plants by 29% and 31%, respectively (Leakey et al. 2009). Experiments have been performed to evaluate the impact of elevated CO_2 and O_3 on crop plants and a variety of responses were observed (Booker et al. 2009; Burkey et al. 2007; Mishra et al. 2013a, b). The most convincing reason to explain the stimulatory effect of CO_2 on O_3 stressed plants is that elevated CO_2 lowers O_3 flux into the leaves due to decline in stomatal conductance, thus restricting the O_3 injury (Kim et al. 2010; Cardoso-Vilhena et al. 2004). As such, the main mechanism underlying the protection incurred by elevated CO_2 is based on the exclusion of O_3 from accumulating in the leaves' apoplast. Cardoso-Vilhena et al. (2004) working on wheat (*T. aestivum* L. cv Hanno) reported that cumulative O_3 exposure reduced by 10% and 35% in 4th and 7th leaf at elevated CO_2 concentration of approximately 700 ppm. In spite of the reduced stomatal conductance, a positive response in the photosynthetic rate was recorded in wheat exposed to elevated CO_2 (Cardoso-Vilhena et al. 2004). Improved photosynthate production in plants upon CO_2 enrichment acts as a adaptive feature that facilitates the plants to compensate for the damage caused under stress conditions. Increased photosynthate availability could be utilized for damage repair and detoxification processes (Mishra et al. 2013a). Under the elevated CO_2 scenario, the intrinsic limitation of photosynthesis shifts from CO_2 fixation in carboxylation towards energy capture by photochemical component of photosynthesis (Long and Drake 1991).

Mishra et al. (2013a) studied the response of two wheat cultivars HUW- 37 and K-9107 treated with elevated O_3 (EO; 58.3 ppb) and elevated CO_2 (EC; 548.2 ppm), singly and in combination (ECO) and observed that elevated CO_2 treatment was able to partially mitigate the negative response of photosynthesis rate in plants exposed to EC and ECO treatments as compared to EO treatment. Rate of photosynthesis reduced by 13% and 10.9% in EO treated wheat cultivars HUW- 37 and K- 9107, respectively as compared to the non filtered plants (NFCs) which served as control. In EC treatment, increments of 23.7% and 31.1% were recorded in HUW-37 and K- 9107, respectively as compared to the control. Under ECO treatment, the response pattern of the different parameters of plants was similar to that of EC treatment, however, the magnitude of response in ECO treated plants was less than EC as compared to control (Mishra et al. 2013a).

Phothi et al. (2016) studied the effect of O_3 (40 and 70 ppb) and elevated CO_2 (700 ppm) on rice (*Oryza sativa* L. cv Khao Dwak Mali 105), singly and in combination and recorded that photosynthetic rate reduced by 25.9% and 49.5%, respectively, in 40 and 70 ppb O_3 treated plants and increased by 61.4%, 49% and 30%, respectively, in plants treated with elevated CO_2 singly and with a combination of elevated CO_2 with 40 and 70 ppb O_3, respectively. Studies have shown that the observed trend of the rate of photosynthesis under different treatments can be explained by the response of the photosynthetic pigments of the experimental plants (Phothi et al. 2016; Mishra et al. 2013a). Chlorophyll and carotenoid contents decreased upon EO treatment, while significantly increased under EC and ECO treatments (Mishra et al. 2013a). Stomatal conductance was found to decrease significantly in all the treatments with the magnitude of reduction being maximum in

ECO treatments (Mishra et al. 2013a). Increased photosynthetic rate in EC and ECO plants despite of reduced stomatal conductance can be explained by CO_2 fertilization, which provided extra carbon for photosynthesis (as discussed earlier).

Broberg (2015) reported that the response of plants under combined treatment with elevated O_3 and CO_2 was of lower magnitude as compared to the response under elevated O_3, suggesting that the negative effect of O_3 is stronger than the positive effect of elevated CO_2, as such CO_2 fertilization was not effective in the complete mitigation of O_3 injury in plants. Liu et al. (2016) reported an initial increase in the rate of photosynthesis followed by a decline at a later growth stage of *Pinus tabulaformis* Carr exposed to a combination of elevated O_3 (80 ppb) and elevated CO_2 (700 ppm). It has been suggested that non stomatal factors such as deactivation of RuBisCO and inhibitory effect on electron transport system may be responsible for reduced rate of photosynthesis during later growth stages of plants (Niu et al. 2014).

The level of antioxidant enzymes such as SOD, POD, CAT, APX and GR reduced in plants treated with elevated CO_2 as compared to the plants growing in ambient CO_2 (Kumari et al. 2013). Reduction in the level of antioxidant enzymes signifies lesser generation of ROS under elevated CO_2 as compared to ambient CO_2. This can be attributed to the fact that elevated CO_2 checked the entry of pollutants, specifically O_3 by reducing the stomatal conductance, thus decreasing the ROS formation. When elevated CO_2 and elevated O_3 were combined together, the enzyme activity increased, but the magnitude of increase was less than that of elevated O_3 alone (Mishra et al. 2013a). This suggests that CO_2 enrichment could partially mitigate O_3 induced damage in plants. However, Kumari et al. (2013) reported enhancement in the levels of antioxidant enzymes in palak (*Beta vulgaris* L. var. Allgreen) exposed to 570 ppm CO_2. This increase was lesser as compared to the increments observed in elevated O_3 and in combination with elevated O_3 and elevated CO_2 interaction treatment (Kumari et al. 2013). Similar observations were also recorded in potato (*Solanum tuberosum* L. cv kufri chandramukhi) exposed to a combination of different concentrations of CO_2 and O_3 viz., ambient CO_2 (ACO_2; 382 ppm), elevated CO_2 (ECO_2; 570 ppm), ambient O_3 (AO_3; 50 ppb) and elevated O_3 (EO_3; 70 ppb) (Kumari and Agrawal 2014). Results of this experiment have shown that the level of injury in plants was directly correlated to O_3 flux under different combinations of ambient and elevated CO_2 and O_3. Maximum injury as determined through lipid peroxidation and solute leakage was found in plants treated with ACO_2 + EO_3 treatment. O_3 stimulated the activities of antioxidative enzymes like SOD, GR and APX in ACO_2 + AO_3 and ACO_2 + EO_3; however elevated CO_2 treatment partially alleviated the alterations of the enzyme activities caused by O_3 stress (Kumari and Agrawal 2014). Secondary metabolites like phenol increased significantly under elevated CO_2 treatment which can be attributed to the allocation of more carbon produced (due to CO_2 fertilization) towards the formation of secondary compounds (Ibrahim et al. 2014). Although the concentration of secondary metabolites in plants increase under O_3 induced oxidative stress, an interaction between elevated CO_2 and elevated O_3 resulted in a further increase in the level of secondary metabolites (Mishra et al. 2013a).

Increase in yield /yield parameters in plants treated with elevated CO_2 is attributed to enhanced assimilate production, increased dry matter accumulation and its allocation towards different yield components such as number of tillers and ears and consequently higher number of grains (Jablonski et al. 2002). Degener (2015) studied the effect of CO_2 fertilization on yield response of 10 crops (wheat, spring wheat, barley, rye, triticale, early, medium and late maize varieties, sunflower and sorghum) in Northern Germany, using a numerical carbon based crop model (BioSTAR). Through this model, yield estimates were done for the entire 21st century, once with changing CO_2 concentration in accordance with SRES-A1B scenario of IPCC (varying from 370 ppm in the year 2000 to 680 ppm in 2090) and secondly with a fixed CO_2 concentration of 390 ppm (Degener 2015). Results of this study by Degener (2015) indicate that generally the yield of the crops increases when modeling was done with changing (rising) CO_2 scenario and decreases when CO_2 is kept constant. Difference in yield response in crops under the two different scenarios (changing and constant CO_2) can be as high as 60% towards the end of the century. In the first half of the century (2011–2030), modeling with increasing CO_2 will increase the average yield of the test plants by 1.3% with barley showing maximum increment of 5%, In the second half of the century, yield increment under the same scenario was 9.7%, with yield increase of 13%, 21% and 22% in barley, sorghum and late maize variety, while the other test crops were confined below 5% yield increment (Degener 2015). On the other hand, under fixed CO_2 scenario, all the crops showed a decline in the yield which was calculated to be $(-)$ 3.6% for the first half of the century and $(-)$ 16.4 for the second half. In this scenario, barley was observed to be most resistant, showing least yield reductions while winter grains, medium maize variety and sunflower were found to be sensitive and were predicted to suffer dramatically if the future CO_2 fertilization was neglected (Degener 2015).

Mishra et al. (2013b) studied the interactive effect of elevated CO_2 and elevated O_3 on yield response of two wheat cultivars (HUW-37 and K-9107) and observed the positive effect of elevated CO_2 in partially mitigating O_3 stress. It was also observed that yield improvement in plants under elevated CO_2 was more prominent in O_3 tolerant cultivar of wheat (K-9107) as compared to the O_3 sensitive one (HUW-36) under similar O_3 and CO_2 exposure conditions (Mishra et al. 2013b). This differential response of different wheat cultivars can be explained by the biomass allocation strategy adopted by the plants in response to O_3 and CO_2 exposure. A comparison of the Harvest index (HI) which determines the partitioning of dry matter between grains (economic yield) and above ground biomass (biological yield) have shown that its value is higher for O_3 resistant K-9107 cultivar as compared to O_3 sensitive HUW- 37 upon treatment with elevated O_3 (Mishra et al. 2013b). This suggests that wheat cultivar K-9107 was able to translocate more of its photosynthate towards the reproductive parts, thus sustaining higher yield. HUW- 37 on the other hand, did not show any significant yield improvement (no effect on HI) under elevated CO_2 indicating that the biomass was equally distributed between vegetative and reproductive parts (Mishra et al. 2013b). Heagle et al. (2000), however, reported higher yield increments in O_3 sensitive wheat C-9904 as compared to O_3 resistant

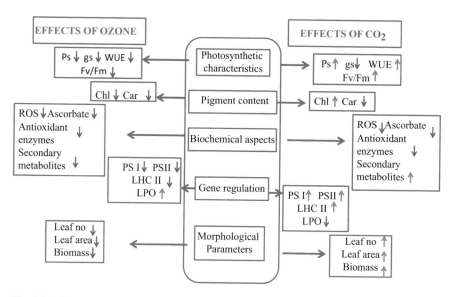

Fig. 5.3 Effect of O$_3$ and CO$_2$ on different parameters of plants

C- 9835 under elevated CO$_2$. Figure 5.3 depicts the response of different parameters of plants under O$_3$ and CO$_2$ treatments.

An important aspect of CO$_2$ fertilization is the differential response of C$_3$ and C$_4$ plants to increasing CO$_2$ concentration (Hager et al. 2016; Kant et al. 2012). The variability in the response of C$_3$ and C$_4$ plants is attributed to the shift in carboxylation/oxygenation efficiency of RuBisCO (Kajala et al. 2011). At low CO$_2$, photosynthesis is limited by the amount of available carbon, therefore the carboxylation activity of RuBisCO is more preferred (Temme et al. 2015). However, under these conditions, the phenomenon of photorespiration becomes prominent and acts as a bottleneck, preventing C$_3$ plants from achieving full photosynthetic potential due to the competition between CO$_2$ and O$_2$ at the C-fixation site of RuBisCO enzyme (Kant et al. 2012). Therefore, with the increasing CO$_2$ concentrations in the atmosphere in near future, the photosynthetic efficiency of the C$_3$ plants is expected to increase due to more C availability thus enhancing the carboxylation efficiency of RuBisCO (Kumar 2016). In C$_4$ plants, the existing carbon concentrating mechanism (CCM) already concentrates 12–20 times CO$_2$ at the C-fixation site of RuBisCO and the carboxylation efficiency of the enzymes is nearly at its saturation limit (Kant et al. 2012; Ainsworth and Rogers 2007). Since the current CO$_2$ levels are sufficient enough to saturate the carboxylation function of RuBisCO in C$_4$ plants, further increments in CO$_2$ level does not affect the productivity of C$_4$ plants (Hager et al. 2016).

Robinson et al. (2012) reported that out of 365 C$_3$ and 37 C$_4$ plant's response to elevated CO$_2$, plant biomass significantly increased in C$_3$ species, but was unchanged in C$_4$ species. This differential response of C$_3$ and C$_4$ plants towards elevated CO$_2$

assumes more significance when the interaction of elevated CO_2 and elevated O_3 is studied. C_4 plants are more dominant in tropical and subtropical areas (Cerling et al. 1993), which are characterized by high temperature, high sunlight intensity and long sunshine hours, which provide more favourable conditions for O_3 formation (Monks et al. 2015; Tiwari et al. 2008). Since C_4 plants are not very responsive to the positive effects of CO_2 fertilization, the negative effects of increasing O_3 concentration will be difficult to check as compared to C_3 plants, which are more responsive to elevated CO_2 (Hager et al. 2016). However, long term exposure of C_3 plants to elevated CO_2 leads to photosynthetic acclimation or down regulation of photosynthetic capacity, depending upon the plant species, developmental stage and environmental conditions (Urban et al. 2012; Sanz-Saez et al. 2013). This can be attributed to the excessive carbohydrate accumulation under elevated CO_2 (Teng et al. 2009), which results in feedback inhibition or physical damage at chloroplast level, reducing the photosynthetic capacity (Aranjuelo et al. 2011).

(b) Nutrient Amendments

Nutrient amendments in soil play an important role in maintaining the sustainability of agricultural production via replacing the nutrients that are removed from the soil by the crops. There are several methods to ensure the replacement of soil nutrients, important ones being biological nitrogen fixation, addition through organic resources like crop residues, livestock manure etc. and commercially used inorganic fertilizers. Nutrients such as nitrogen (N), phosphorous (P) and potassium (K) are the most important in defining crop productivity and yield and are most commonly supplemented to maintain/enhance the agricultural production (Singh et al. 2015). Of these, nitrogen is an important nutrient required for plant growth and development as it is the core constituent of plant's nucleic acid, enzymes and cell wall pigment system (Krapp 2015). Plants frequently exposed to N stress conditions show limited growth (Iqbal et al. 2015). Higher N supply contributes to delayed leaf senescence in O_3 stressed plants by maintaining higher protein levels (Zeng et al. 2017; Diaz-Mendoza et al. 2016). Higher N content remobilizes carbon from leaves to the reproductive parts before senescence to maintain yield under stress conditions (Distelfeld et al. 2014).

Potassium plays a significant role in transport of assimilates and activates the enzymes of assimilate metabolism and their conversion to oils (Wang et al. 2013). Potassium regulates NADPH activity which reduces the ROS generated during stress (Wang et al. 2013). Reduced ROS in plants due to potassium amendments leads to cell membrane stability and integrity, thereby protecting chlorophyll structure in the chloroplast (Wang et al. 2013). Potassium regulates stomatal conductance which is known to decline under O_3 stress. In addition to this potassium promotes root and shoot growth, leaf area and total dry mass of plants under abiotic stress (Wang et al. 2013).

Interactive studies of O_3 stress and nutrient amendments have shown that nutrients are effective in bringing about partial mitigation of O_3 injury by increasing the photosynthetic rate due to higher availability of nitrogen, which is a crucial factor in photosynthesis (Singh et al. 2009). High nutrient availability increased the photo-

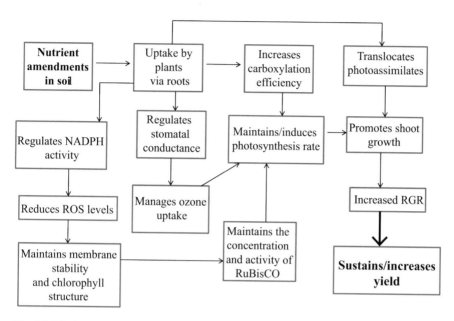

Fig. 5.4 Mechanism of action of nutrient amendments in plants

synthetic carbon fixation even under O₃ stress and thus helped in increasing the yield of the plants (Singh et al. 2011). Yamaguchi et al. (2007a) have shown that in *Fagus crenata* seedlings, the interactive effect of O₃ and nitrogen interaction does not affect the stomatal conductance, therefore the observed modification in the photosynthetic rate could not be explained by stomatal limitation to photosynthetic CO₂ fixation. Rather, the increased photosynthetic rate was attributed to increased carboxylation efficiency and concentration of RuBisCO in O₃ and nitrogen treated plants (Yamaguchi et al. 2007a). Although the whole plant biomass did not show any significant change during the first year of growth, at the end of the second growing season nitrogen amendments induced significant increase in whole plant biomass of *F. crenata* seedlings (Yamaguchi et al. 2007b). These observations suggest that the nitrogen supplied to the soil in the previous year strongly affects the whole plant growth in the next season (Yamaguchi et al. 2007b). Figure 5.4 represents the effect of nutrient amendments on different physiological and biochemical events and its effect on the yield of O₃ stressed plants.

Nutrient amendments also modify the morphological response of O₃ stressed plants. Singh et al. (2015) treated two cultivars of wheat (HUW- 510 and LOK-1) growing at O₃ concentrations ranging from 10.3–110 ppb (12 h) with two doses of NPK [recommended (NPK) and 1.5 times recommended (1.5 NPK)] and found that 1.5 NPK dose decreased root length significantly in HUW-510, whereas the same cultivar showed increments in shoot length as compared to other treatments. This indicates that under higher availability of nutrients, root growth is modified negatively as compared to lower fertilization dose (Singh et al. 2015). Increased shoot length accompanied by reduction in root length at higher nutrient availability (1.5 NPK) as compared to lower nutrient availability (NPK) suggests that the nutrients facilitate the allocation of more assimilate towards the above ground parts as compared to the below ground parts under similar O_3 exposure regimes (Singh et al. 2015). RGR increased significantly in O_3 stressed plants as compared to control (ambient O_3) in HUW-510 under recommended NPK. In LOK-1, RGR showed increments at 1.5NPK; however the value of NAR showed a significant reduction. In this experiment, increment in RGR is explained by an increase is LAR which is large enough to override the observed reduction in NAR (Singh et al. 2015). A comparison between NAR values of the two cultivars indicates that HUW-510 was more efficient than LOK-1 in fixing carbon at recommended NPK (Singh et al. 2015). The yield reductions under O_3 and nutrient interaction treatments also suggest that HUW- 510 is a tolerant variety than LOK-1 (Singh et al. 2015).

2.2   Adapting Crops to Ozone Exposure

(a) Screening of O_3 Tolerant Cultivars

In addition to improving the air quality and modifying plant's response through CO_2 fertilization and nutrient amendments, another strategy that can be used for mitigation of O_3 induced injury in plants is to focus on adapting crops to O_3 exposure. This can be done through screening O_3 resistant cultivars and promoting their use in agricultural practices in areas prone to high risks of O_3. Large scale field studies conducted in United States and Europe in 1980/1990s established the existence of a wide range of crop sensitivity to O_3, both among different crops and between different cultivars of the same crop (Heagle 1989; Heck 1989; Krupa et al. 1998). Along with these large scale experimental studies, some individual experiments done to evaluate the impact assessment of O_3 have also shown a prominent variation in the sensitivity of different crops/cultivars to O_3 stress (Singh et al. 2010; Saitanis et al. 2014; Manigbas et al. 2013).

Various workers have reported that the O_3 sensitivity of the crops/cultivars might be more genotype dependant and can vary by as much as 30% across the cultivars (Biswas et al. 2008; Gonzalez-Fernandez et al. 2010). Monga et al. (2015) studied the relative sensitivities of 5 Mediterranean winter wheat cultivars (Sculptur, Colombo, Pharaon, Gallareta and Vitron) grown under similar O_3 exposure conditions and observed that the physiological response of wheat to O_3 exposure was cultivar dependant. On the basis of above ground biomass, Sculptur and Colombo were found to be sensitive cultivars, however, significant yield reductions were recorded only in Sculptur. Vitron behaved as a moderately sensitive cultivar which showed significant reductions in stem dry biomass and number of filled and empty spikes, which altogether resulted in no significant reduction in grain yield (Monga et al. 2015). The other two cultivars, Gillareta and Pharaon did not show any significant variations in biomass or yield when exposed to O_3 and were categorized as O_3

resistant (Monga et al. 2015). In this experiment, biomass allocation towards reproductive development played an important role in defining the sensitivity of wheat cultivars Colombo and Vitron. Biomass allocation to reproductory parts is a compensatory mechanism to improve the reproductive output in response to O_3 stress (Black et al. 2012; Gerosa et al. 2009).

Ozone sensitivity can also be related to physiological traits such as stomatal conductance, photosynthetic capacity and detoxification ability of the crop/cultivar (Biswas et al. 2008). In the experiment conducted by Monga et al. (2015) sensitive cultivars Sculptur and Colombo showed higher levels of stomatal conductance than the tolerant cultivars Gallareta and Pharon. However, a few researchers failed to find a clear relationship between stomatal conductance and O_3 sensitivity (Gonzalez-Fernandez et al. 2010) suggesting that certain other factors also determine intra species changes in O_3 sensitivity (Feng et al. 2016).

Singh et al. (2010) reported soybean (*Glycine max* L.) cultivar PK472 to be O_3 sensitive and cultivar Bragg to be O_3 resistant on the basis of their calculated yield response to stress (YRS) grown under similar regimes of elevated O_3 (70 and 100 ppb). An important feature of this study was that although the reduction in photosynthetic rate was higher in case of Bragg (25.6% and 32% at O_3 concentration of 70 and 100 ppb, respectively) than in PK472 (19.8% and 40.4% at O_3 concentration of 70 and 100 ppb, respectively) (Singh et al. 2009), yield reductions were higher in PK472 (20% and 33% at O_3 concentration of 70 and 100 ppb, respectively) than in Bragg (12% and 30% at O_3 concentration of 70 and 100 ppb, respectively) (Singh et al. 2010). This study indicated that in spite of higher reductions in photosynthetic rate, Bragg was able to sustain lower reductions in yield. PK472, on the other hand was not able to uphold its yield despite of higher photosynthetic yield. This unpredicted result can be explained by the biomass allocation pattern of the two soybean cultivars. It was observed that Bragg showed a tendency of translocating more of its biomass towards the reproductive structures (pods), contrary to PK472 which showed a normal distribution of biomass between reproductive and vegetative parts of the plant (Singh et al. 2010). This study emphasizes the role of biomass allocation in defining the cultivar sensitivity to O_3 stress.

A few other studies have also investigated the cultivar sensitivity towards O_3 stress. Saitanis et al. (2014) screened the Bangladeshi winter wheat cultivars for their sensitivity to O_3 on the basis of visible injury and chl a/b ratio. Ten wheat cultivars were exposed to 8 O_3 exposure regimes (50, 60, 80, 100, 120, 135, 150 and 200 ppb for 14, 11, 8, 6, 5, 4, 3 and 1 days, respectively for 8hd⁻¹) in controlled environmental chambers and observed the order of sensitivity to be Akbar >> Sufi > Bijoy > Shatabdi > Bari- 26 > Gaurab >Bari- 25 > Prodip > Saurav >> Kanikan.

Avnery et al. (2013) estimated the crop production loss variation by 2030 by selecting soybean, maize and wheat cultivars with demonstrated tolerance to O_3 relative to baseline crop sensitivity. It was observed that total (soybean, maize and wheat) year 2030 crop production loss (CPL) was 81 Mt., when cultivars with greatest tolerance to O_3 stress were used and was 143 Mt. with sensitive cultivars (Avnery et al. 2013). This is equivalent to a ~12% improvement from the year 2000 production. The crop production (CP) gains were highest for wheat, representing the prevention of

64% of CPL, otherwise projected to occur in 2030 (Avnery et al. 2013). By choosing most tolerant cultivars of maize and soybean, CP improved by 1.6% and 8%, respectively relative to the year 2000, representing 55–76% reductions in the losses expected to occur by 2030 with O_3 sensitive cultivars (Avnery et al. 2013). The CP gains through the selection of tolerant cultivars were expected to be maximum (90% of the regional production in 2000) in the Indian subcontinent where the rise in O_3 is predicted to be greatest between 2005 and 2030. It was observed that India and Pakistan would acquire the greatest economic benefits from increased selection of O_3 tolerant cultivars (Avnery et al. 2013).

(b) Improving Photosynthetic Efficiency of Plants

Rate of photosynthesis per unit leaf area is an important parameter in determining the plant productivity. As discussed earlier, C_4 plants are already at their saturation level of C fixation and there is not much room left for increasing their photosynthetic rate (Kant et al. 2012; Ainsworth and Rogers 2007). C_3 plants, however, do show a scope for increasing the rate of photosynthesis as their C fixation is limited by low concentration of CO_2 at the active sites of RuBisCO (Kumar 2016). Several approaches have been adopted for improving the photosynthetic rate in C_3 plants, a few of them are:

(i) Introducing C_4 like characteristics in C_3 cells (Kajala et al. 2011; Miyao et al. 2011).
(ii) Introducing CO_2 pump proteins into chloroplast membrane from cyanobacteria (Price et al. 2008)
(iii) To overcome photorespiration by introducing new catabolic pathways into plastids (Kebeish et al. 2007).
(iv) Improving RuBisCO kinetics characteristics.

Studies have shown that C_4 photosynthesis has evolved independently in 19 different families of angiosperms (Sage et al. 2011) and there are at least 21 examples displaying intermediate C_3-C_4 photosynthetic characteristics (Edwards et al. 2004). It is important to note that some of the enzymes and structures present in C_4 plants are also found in C_3 plants (Ehleringer et al. 1991). This fact is used by plant biologists attempting to engineer C_4 pathway in C_3 plants through identification of genotypes expressing some degree of cellular similarities to C_4 plants (Furbank et al. 2009). Partial C_4 cycle genes have been introduced into rice, tobacco and potato (Kajala et al. 2011). It is difficult to engineer kranz anatomy in C_3 plants as it requires targeting multiple genes (Kajala et al. 2011). However, two species of the family chenopodiaceae exhibit C_4 photosynthetic system contained within a single collenchyma cell performing the function of kranz anatomy by concentrating CO_2 around RuBisCO (Edwards et al. 2004), which gives a hope of successful introduction of C_4 cycle in C_3 plants in a near future. Attempts are been made to achieve this target in rice by expression of specific C_4 cycle genes in rice (Kajala et al. 2011; Miyao et al. 2011). However, a functional C_4 cycle in C_3 species is yet not achieved.

Another important technique of increasing photosynthetic efficiency of C_3 plants is to introduce the cyanobacterial CO_2 concentration mechanism in C_3 plants. As a photosynthetic organism, cyanobacteria have evolved a very efficient mechanism for converting CO_2 into HCO^{3-}, an important step in transporting C into the chloroplast stroma. There have been a number of attempts made to introduce CCM into C_3 plants, but so far limited progress has been achieved. A cyanobacterial gene *ict*B, which is responsible for CO_2 accumulation in cyanobacteria, when introduced in *Arabidopsis* and tobacco resulted in significant increase in photosynthetic rates (Lieman-Hurwitz et al. 2003). These studies indicate that introduction of C_4 cycle characteristics in C_3 plants is well within the scope of modern genetic technology and can give fruitful results in near future.

3 Conclusion

Earlier works have well quantified the present and potential future (2030) impacts of surface O_3 on the global yield of soybean, maize and wheat, which are the important food crops of the world. It has been predicted that substantial future yield losses for these crops are expected in near future even under the condition of stringent O_3 control via traditional pollution mitigation measures. This chapter has discussed the different strategies utilized for maintaining yield losses in O_3 stressed plants. Air quality management via targeting traditional O_3 precursors like CH_4 and NOx, modification of plant response via different agronomic practices like CO_2 fertilization and nutrient amendments and adaptational strategies like use of resistant cultivars are some of the techniques that can be utilized for the partial mitigation of O_3 injury in plants. It has been observed that the use of O_3 tolerant cultivars would produce high potential agricultural benefits as compared to O_3 abatement through the control of CH_4 emissions. However, cultivar selection is an individualistic approach and needs to be implemented in each crop seperately. Therefore CH_4 emission control is considered to be more effective strategy for long term international air quality management.

CO_2 fertilization and nutrient amendments bring about modifications in plants' response to O_3 stress. These strategies protect the yield of the plants and minimize the O_3 induced yield reductions. The main target of these two strategies is to increase the photosynthetic rate which allows the incorporation of more carbon in the plant biomass, ultimately assisting in sustained plant yield even under O_3 stress conditions. Although considerable uncertainties remain, O_3 mitigation strategies discussed in this chapter provide important opportunities to significantly improve crop production. Observable partial mitigation of O_3 stress through adoption of the discussed strategies may play an important role in alleviating to some extent, the threat to global food security in near future.

References

AbdElgawad H, Zinta G, Beemster GTS, Janssens IA, Asard H (2016) Future climate CO_2 level mitigate stress impact on plants: increased defense or decreased challenge? Front Plant Sci 7:556–562

Ainsworth EA, Rogers A (2007) The response of photosynthesis and stomatal conductance to rising CO_2: mechanisms and environmental interactions. Plant Cell Environ. 30:258–270

Alexandratos N, Bruinsma J (2012) World agriculture towards 2030/2050: the 2012 revision (Vol. 12). ESA Working Paper

Aranjuelo I, Cabrera-Bosquet L, Morcuende R, Avice JC, Nogues S, Araus JL, Martinez-Carrasco R, Perez P (2011) Does ear C sink strength contribute to overcoming photosynthetic acclimation of wheat plants exposed to elevated CO_2? J Exp Bot 62:3957–3969

Avnery S, Mauzerall DL, Liu J, Horowitz LW (2011a) Global crop yield reductions due to surface ozone exposure: 1. year 2000 crop production losses and economic damage. Atmos Environ 45:2284–2296

Avnery S, Mauzerall DL, Liu J, Horowitz LW (2011b) Global crop yield reductions due to surface ozone exposure: 2 year 2030 potential crop production losses and economic damage under two scenarios of O_3 pollution. Atmos Environ 45:2297–2309

Avnery S, Mauzerall DL, Fiore AM (2013) Increasing global agricultural production by reducing ozone damages via methane emission controls and ozone resistant cultivar selection. Glob Chang Biol 19:1285–1299

Biswas DK, Xu H, Li YG, Liu MZ, Chen YH, Sun JZ, Jiang GM (2008) Assessing the genetic relatedness of higher ozone sensitivity of modern wheat to its wild and cultivated progenitors/relatives. J Exp Bot 59:951–963

Black VJ, Stewart CA, Roberts JA, Black CR (2012) Timing of exposure to ozone affects reproductive sensitivity and compensatory ability in *Brassica campestris*. Environ Exp Bot 75:225–234

Booker F, Muntifering R, McGrath M, Burkey K, Decoteau D, Fiscus E, Manning WJ, Krupa S, Chappelka A, Grantz D (2009) The ozone component of global change: potential effects on agricultural and horticultural plant yield, product quality and interactions with invasive species. J Integr Plant Biol 51:337–351

Bortolin RC, Caregnato FF, Divan AM Jr, Reginatto FH, Gelain DP, Moreira JC (2014) Effects of chronic elevated ozone concentration on the redox state and fruit yield of red pepper plant *Capsicum baccatum*. Ecotoxicol Environ Saf 100:114–121

Broberg M (2015) Effects of elevated ozone and carbon dioxide on wheat crop yield – meta-analysis and exposure-response relationships. Degree project for masters of science. University of Gothenberg, Gothenburg

Bunce JA (2014) Limitations to soybean photosynthesis at elevated carbon dioxide in free-air enrichment and open top chamber systems. Plant Sci 226:131–135

Burkey KO, Booker FL, Pursley WA, Heagle AS (2007) Elevated carbon dioxide and ozone effects on peanut. II. Seed yield and quality. Crop Sci 47:1488–1497

Cardoso-Vilhena J, Balaguer L, Eamus D, Ollerenshaw J, Barnes J (2004) Mechanisms underlying the amelioration of O_3-induced damage by elevated atmospheric concentrations of CO_2. J Exp Bot 55(397):771–781

Cerling TE, Wang Y, Quade J (1993) Expansion of C4 ecosystems as an indicator of global ecological change in late Miocene. Nature 361:344–345

Cofala J, Amann M, Klimont Z, Kupiainen K, Höglund-Isaksson L (2007) Scenarios of global anthropogenic emissions of air pollutants and methane until 2030. Atmos Environ 41:8486–8499

Danh NT, Huy LH, Oanh NTK (2016) Assessment of rice yield loss due to exposure to ozone pollution in Southern Vietnam. Sci Total Environ 566-567:1069–1079

Degener JF (2015) Atmospheric CO_2 fertilization effects on biomass yield of 10 crops in northern Germany. Front Environ Sci. 3:48–61

Diaz-Mendoza M, Velasco-Arroyo B, Santamaria ME, González-Melendi P, Martinez M, Diaz I (2016) Plant senescence and proteolysis: two processes with one destiny. Genet Mol Biol 39(3):329–338

Distelfeld A, Avni R, Fischer AM (2014) Senescence, nutrient remobilization, and yield in wheat and barley. J Exp Bot 65:3783–3798

Edwards GE, Franceschi VR, Voznesenskaya EV (2004) Single cell C_4 photosynthesis versus the dual cell (Kranz) paradigm. Annu Rev Plant Biol 55:173–196

Ehleringer JR, Sage RF, Flanagan LB, Pearcy RW (1991) Climate change and the evolution of C_4 photosynthesis. Trends Ecol Evol. 6:95–99

Farfan-Vignolo ER, Asard H (2012) Effect of elevated CO_2 and temperature on the oxidative stress response to drought in *Lolium perenne* L. and *Medicagosativa* L. Plant Physiol Biochem 59:55–62

Feng Z, Hu E, Wang X, Jiang L, Liu X (2015) Ground-level O_3 pollution and its impacts on food crops in China: a review. Environ Pollut 199:42–48

Feng Z, Wang L, Pleijel H, Zhu J, Kobayashi K (2016) Differential effects of ozone on photosynthesis of winter wheat among cultivars depend on antioxidative enzymes rather than stomatal conductance. Sci Total Environ 572:404–411

Fiore AM, West JJ, Horowitz LW, Naik V, Schwarzkopf DM (2008) Characterizing the tropospheric ozone response to methane emission controls and the benefits to climate and air quality. J Geophys Res 113:D08307

Forster P, Ramaswamy V, Artaxo P et al (2007) Contribution of working group I to the fourth assessment report of the intergovernmental panel on climate change. In: Solomon S, Qin D, Manning M, Chen Z, Marquis M, Averyt KB, Tignor M, Miller HL (eds) Climate change 2007: the physical science basis. Cambridge University Press, Cambridge/New York, pp 129–234

Furbank RT, vonCaemmerer S, Sheehy J, Edwards G (2009) C_4 rice: a challenge for plant phenomics. Funct Plant Biol 36:845–856

Gerosa G, Marzuoli R, Rossini M, Panigada C, Meroni M, Colombo R, Faoro F, Iriti M (2009) A flux-based assessment of the effects of ozone on foliar injury, photosynthesis, and yield of bean (*Phaseolus vulgaris* L. cv. Borlotto Nano Lingua di Fuoco) in open-top chambers. Environ Pollut 157:1727–1736

Ghasemzadeh A, Jaafar HZE, Rahmat A (2010) Elevated carbondioxide increases contents of flavonoids and phenolic compounds and antioxidant activities in Malaysian young ginger (*Zingiber officinale* roscoe.) varieties. Molecules 15:7907–7922

Ghude SD, Jena CK, Beig G, Kumar R, Kulkarni SH, Chate DM (2016) Impact of emission mitigation on ozone-induced wheat and rice damage in India. Curr Sci 110(8):1452–1458

Gill SS, Tuteja N (2010) Reactive oxygen species and antioxidant machinery in abiotic stress tolerance in crop plants. Plant Physiol Biochem 48:909–930

Gonzalez-Fernandez I, Kaminska A, Dodmani M, Goumenaki E, Quarrie S, Barnes JD (2010) Establishing ozone flux-response relationships for winter wheat: analysis of uncertainties based on data for UK and polish genotypes. Atmos Environ 44:621–630

Guo H, Sun Y, Li Y, Tong B, Herris M, Zhu-Salzman K et al (2013) Pea aphid promotes amino acid metabolism both in *Medicago truncatula* and bacteriocytes to favor aphid population growth under elevated CO_2. Glob Chang Biol 19:3210–3223

Hager HA, Ryan GD, Kovacs HM, Newman JA (2016) Effects of elevated CO_2 on photosynthetic traits of native and invasive C_3 and C_4 grasses. BMC Ecol 16:28–40

Heagle AS (1989) Ozone and crop yield. Annu Rev Phytopathol 27:397–423

Heck WW (1989) Assessment of crop losses from air pollutants in the United States. In: MacKenzie JJ, El-Ashry MT (eds) Air pollution's toll on forests and crops. Yale University Press, New Haven, pp 235–315

Heagle AS, Miller JE, Pursley WA (2000) Growth and Yield Responses of Winter Wheat to Mixtures of Ozone and Carbon Dioxide. Crop Sci 40:1656–1664

Hodges DM, Forney CF (2000) The effects of ethylene, depressed oxygen and elevated carbon dioxide on antioxidant profiles of senescing spinach leaves. J Exp Bot 51:645–655

Hogy P, Brunnbauer M, Koehler P, Schwadorf K, Breuer J, Franzaring J, Zhunusbayeva D, Fangmeier A (2013) Grain quality characteristics of spring wheat (*Triticum aestivum*) as affected by free-air CO_2 enrichment. Environ Exp Bot 88:11–18

Horowitz LW, Walters S, Mauzerall DL et al (2003) A global simulation of tropospheric ozone and related tracers: description and evaluation of MOZART, Version 2. J Geophys Res 108:4784. https://doi.org/10.1029/2002JD002853

IPCC (2013) Summary for policymakers. In: Stocker TF, Qin D, Plattner G-K, Tignor M, Allen SK, Bouschung J, Nauels A, Xia Y, Bex V, Midgley PM (eds) Climate change 2013: the physical science basis. Contribution of working group I to the fifth assessment report of intergovermental panel on climate change. Camebridge University Press, Cambridge/New York

Ibrahim MH, Jaafar HZE, Karimi E, Ghasemzadeh A (2014) Allocation of secondary metabolites, photosynthetic capacity, and antioxidant activity of Kacip Fatimah (Labisia pumila Benth) in response to CO_2 and light intensity. Sci World J 2014:360290–360213

Iqbal N, Umar S, Khan NA (2015) Nitrogen availability regulates proline and ethylene production and alleviates salinity stress in mustard (*Brassica juncea*). J Plant Physiol 178:84–91

Jablonski LM, Wang X, Curtis PS (2002) Plant reproduction under elevated CO_2 conditions: a meta-analysis of reports on 79 crop and wild species. New Phytol 156:9–26

Jagadish KSV, Kadam NN, Xiao G, Melgar RJ, Bahuguna RN, Quinones C et al (2014) Agronomic and physiological responses to high temperature, drought, and elevated CO_2 interactions in cereals. AdvAgron 127:111–156. https://doi.org/10.1016/B978-0-12-800131-8.00003-0

Kajala K, Covshoff S, Karki S, Woodfield H, Tolley BJ, Dionora MJA, Mogul RT, Mabilangan AE, Danila FR, Hibberd JM, Quick WP (2011) Strategies for engineering a two-celled C_4 photosynthetic pathway in to rice. J Exp Bot 62:3001–3010

Kant S, Seneweera S, Rodin J, Materne M, Burch D, Rothstein SJ, Spangenberg G (2012) Improving yield potential in crops under elevated CO_2: integrating the photosynthetic and nitrogen utilization efficiencies. Front Plant Sci 3(162):1–9

Kebeish R, Niessen M, Thiruveedhi K, Bari R, Hirsch HJ, Rosenkranz R, Stabler N, Schonfeld B, Kreuzaler F, Peterhansel C (2007) Chloroplastic photorespiratory bypass increases photosynthesis and biomass production in *Arabidopsis thaliana*. Nat Biotechnol 25:593–599

Kim TH, Böhmer M, Hu H, Nishimura N, Schroeder JI (2010) Guard cell signal transduction network: advances in understanding abscisic acid, CO_2, and Ca^{2+} signaling. Annu Rev Plant Biol 61:561–591

Kotchoni SO, Kuhns C, Ditzer A, Kirch HH, Bartels D (2006) Over-expression of different aldehyde dehydrogenase genes in *Arabidopsis thaliana* confers tolerance to abiotic stress and protects plants against lipid peroxidation and oxidative stress. Plant Cell Environ. 29:1033–1048

Krapp A (2015) Plant nitrogen assimilation and its regulation: a complex puzzle with missing pieces. Curr Opin Plant Biol 25:115–122

Krupa SV, Tonneijck AEG, Manning WJ (1998) Ozone. In: Flagler RB (ed) Recognition of air pollution injury to vegetation: a pictorial atlas. Air & Waste Management Association, Pittsburgh, pp 2–28

Kumar M (2016) Impact of climate change on crop yield and role of model for achieving food security. Environ Monit Assess 188:465–478

Kumari S, Agrawal M (2014) Growth, yield and quality attributes of a tropical potato variety (*Solanum tuberosum* L. cv Kufri chandramukhi) under ambient and elevated carbon dioxide and ozone and their interactions. Ecotoxicol Environ Saf 101:146–156

Kumari S, Agrawal M, Tiwari S (2013) Impact of elevated CO_2 and elevated O_3 on *Beta vulgaris* L.: pigments, metabolites, antioxidants, growth and yield. Environ Pollut 174:279–288

Leakey ADB, Ainsworth EA, Bernacchi CJ, Rogers A, Long SP, Ort DR (2009) Elevated CO_2 effects on plant carbon, nitrogen, and water relations: six important lessons from FACE. J Exp Bot 60:2859–2876

Lieman-Hurwitz J, Rachmilevitch S, Mittler R, Marcus Y, Kaplan A (2003) Enhanced photosynthesis and growth of transgenic plants that express *ict*B, a gene involved in HCO_3^- accumulation in cyanobacteria. Plant Biotechnol J 1:43–50

Liu Z, Chen W, Fu W, He X, Fu S, Lu T (2016) Effects of elevated CO_2 and O_3 on leaf area, gas exchange and starch contents in Chinese pine (*Pinus tabulaeformis* Carr) in northern China. Bangladesh J Bot 44(5):917–923

Long SP, Drake BG (1991) Effect of the long –term elevation of CO_2 concentration in the field on the quantum yield of photosynthesis of the C_3 sedge, *Scirpusolneyi*. *Plant Physiol* 96:221–226

Manigbas NL, Park D-S, Park S-K, Kim S-M, Hwang W-H, Kang H-W, Yi G (2013) Development of a fast and reliable ozone screening method in rice (*Oryza sativa* L.) Afr J Crop Sci 1(1):11–17

Markelz RC, Lai LX, Vosseler LN, Leakey AD (2014) Transcriptional reprogramming and stimulation of leaf respiration by elevated CO_2 concentration is diminished, but not eliminated, under limiting nitrogen supply. Plant Cell Environ. 37:886–898

Mishra AK, Agrawal SB (2014) Cultivar specific response of CO_2 fertilization on two tropical mungbean (*Vigna radiata* L.) cultivars: ROS generation, antioxidant status, physiology, growth, yield and seed quality. J Agron Crop Sci. 20:273–289

Mishra AK, Rai R, Agrawal S (2013a) Individual and interactive effects of elevated carbon dioxide and ozone on tropical wheat (*Triticum aestivum* L.) cultivars with special emphasis on ROS generation and activation of antioxidant defence system. Indian J Biochem Biophys 50:139–149

Mishra AK, Rai R, Agrawal SB (2013b) Differential response of dwarf and tall tropical wheat cultivars to elevated ozone with and without carbon dioxide enrichment: Growth, yield and grain quality. Field Crop Res 145:21–32

Miyao M, Masumoto C, Miyazawa SI, Fukayama H (2011) Lessons from engineering a single cell C_4 photosynthetic pathway in to rice. J Exp Bot 62:3021–3029

Monga R, Marzuoli R, Alonso R, Bermejo V, Gonzalez-Fernandez I, Faoro F, Gerosa G (2015) Varietal screening of ozone sensitivity in Mediterranean durum wheat (*Triticum durum* Desf.) Atmos Environ 110:18–26

Monks PS, Archibald TA, Colette A, Cooper O, Coyle M, Derwent R, Fowler D, Granier C, Law KS, Mills GE, Stevenson DS, Tarasova O, Thouret V, von Schneidemesser E, Sommariva R, Wild O, Williams ML (2015) Tropospheric ozone and its precursors from the urban to the global scale from air quality to short-lived climate forcer. Atmos Chem Phys 15:8889–8973

Niu J, Feng Z, Zhang W, Zhao P, Wang X (2014) Non-stomatal limitation to photosynthesis in *Cinnamomum camphoras* seedlings exposed to elevated O_3. PLoS One 9(6):e98572

Osborne SA, Mills G, Hayes F, Ainsworth EA, Buker P, Embersen L (2016) Has the sensitivity of soybean cultivars to ozone pollution increased with time? An analysis of published dose–response data. Glob Chang Biol 22:3097–3111

Peñuelas J, Sardans J, Estiarte M, Ogaya R, Carnicer J, Coll M et al (2013) Evidence of current impact of climate change on life: a walk from genes to the biosphere. Glob Chang Biol 19:2303–2338

Phothi R, Umponstira C, Sarin C, Siriwong W, Nabheerong N (2016) Combining effects of ozone and carbon dioxide application on photosynthesis of Thai jasmine rice (Oryza Sativa L.) cultivar Khao Dawk Mali 105. Aust J Crop Sci 10(4):591–597

Price GD, Badger MR, Woodger FJ, Long BM (2008) Advances in understanding the cyanobacterial CO_2 concentrating mechanism (CCM): functional components, Ci transporters, diversity, genetic regulation and prospects for engineering in to plants. J Exp Bot 59:1441–1461

Ribeiro DM, Araujo WL, Fernie AR, Schippers JHM, Mueller- Roeber B (2012) Action of Gibberellins on growth and metabolism of *Arabidopsis* plants associated with high concentration of carbon dioxide. Plant Physiol 160:1781–1794

Robinson EA, Ryan GD, Newman JA (2012) A meta-analytical review of the effects of elevated CO_2 on plantarthropod interactions highlights the importance of interacting environmental and biological variables. New Phytol 194:321–336

Sage RF, Christin PA, Edwards EJ (2011) The C_4 plant lineages of planet Earth. J Exp Bot 62:3155–3169

Saitanis CJ, Bari SM, Burkey KO, Stamatelopoulos D, Agathokleous E (2014) Screening of Bangladeshi winter wheat (*Triticum aestivum* L.) cultivars for sensitivity to ozone. Environ Sci Pollut Res 21(23):13560–13571

Sanz-Sáez Á, Erice G, Aranjuelo I, Aroca R, Ruíz-Lozano JM, Aguirreolea J et al (2013) Photosynthetic and molecular markers of CO_2- mediated photosynthetic down regulation in nodulated alfalfa. J Integr Plant Biol 55:721–734

Sarkar A, Datta S, Singh P (2017) Tropospheric Ozone Pollution, Agriculture, and Food Security. In: Singh RP, Singh A, Srivastava V (eds) Environmental issues surrounding human overpopulation. IGI Global, Hershey, pp 234–252

Sharma P, Sharma ABJ, Dubey RS, Pessarakli M (2012) Reactive oxygen Species, oxidative damage, and antioxidative defense mechanism in plants under stressful conditions. J Bot 2012:217–237

Shindell D, Kuylenstierna JCI, Vignati E et al (2012) Simultaneously mitigating near-term climate change and improving human health and food security. Science 335:183–189

Singh A, Agrawal M (2015) Effects of ambient and elevated CO_2 in growth, chlorophyll fluorescence, photosynthetic pigments, antioxidants, and secondary metabolites of *Catharanthus roseus* (L.) GDon. grown under three different soil N levels. Environ Sci Pollut Res 22:3936–3946

Singh P, Agrawal M, Agrawal SB (2009) Evaluation of physiological growth and yield responses of a tropical oil crop (*Brassica campestris* L. var. Kranti) under ambient ozone pollution at varying NPK levels. Environ Pollut 157:871–880

Singh E, Tiwari S, Agrawal M (2010) Variability in antioxidant and metabolite levels, growth and yield of two soybean varieties: an assessment of anticipated yield losses under projected elevation of ozone. Agric Ecosyst Environ 135:168–177

Singh P, Agrawal M, Agrawal SB (2011) Differences in ozone sensitivity at different NPK levels of three tropical varieties of mustard (*Brassica campestris* L.): photosynthetic pigments, metabolites, and antioxidants. Water Air Soil Pollut. 214:435–450

Singh P, Agrawal M, Agrawal SB, Singh S, Singh A (2015) Genotypic differences in utilization of nutrients in wheat under ambient ozone concentrations: growth, biomass and yield. Agric Ecosyst Environ 199:26–33

Tai AP, Martin MV, Heald CL (2014) Threat to future global food security from climate change and ozone air pollution. Nat Clim Chang. https://doi.org/10.1038/NCLIMATE2317

Tang H, Takigawa M, Liu G, Zhu J, Kobayashi K (2013) A projection of ozone- induced wheat production loss in China and India for the years 2000 and 2020 with exposure-based and flux-based approaches. Glob Chang Biol 19:2739–2752

Temme AA, Liu JC, Cornwell WK, Cornelissen JHC, Aerts R (2015) Winners always win: growth of a wide range of plant species from low to future high CO_2. Ecol Evol. 5:4949–4961

Teng N, Jin B, Wang Q, Hao H, Ceulemans R, Kuang T et al (2009) No detectable maternal effects of elevated CO_2 on *Arabidopsis thaliana* over 15 generations. PLoS One 4:e6035

Tester M, Langridge P (2010) Breeding technologies to increase crop production in a changing world. Science 327:818–822

Tian H, Ren W, Tao B, Sun G, Chappelka A, Wang X, Pan S, Yang J, Liu J, Felzer BS, Melillo JM, Reilly J (2016) Climate extremes and ozone pollution: a growing threat to China's food security. Ecosyst Health Sustain 2(1):e01203

Tiwari S, Vaish B, Singh P (2017) Population and global food security: issues related to climate change. In: Singh RP, Singh A, Srivastava V (eds) Environmental issues surrounding human overpopulation. IGI Global, Hershey, pp 40–63

Tiwari S, Rai R, Agrawal M (2008) Annual and seasonal variations in tropospheric ozone concentrations around Varanasi. Int J Remote Sens 9(15):4499–4514

Tripathy BC, Oelmüller R (2012) Reactive oxygen species generation and signaling in plants. Plant Signal Behav 7(12):1621–1633

Urban O, Hrstka M, Zitová M, Holišová P, Šprtová M, Klem K et al (2012) Effect of season, needleage and elevated CO_2 concentration on photosynthesis and Rubisco acclimation in *Picea abies*. Plant Physiol Biochem 58:135–141

Van Dingenen R, Raes F, Krol MC, Emberson L, Cofala J (2009) The global impact of O_3 on agricultural crop yields under current and future air quality legislation. Atmos Environ 43:604–618

Wang M, Zheng Q, Shen Q, Guo S (2013) the critical role of potassium in plant stress response. Int J Mol Sci. 14:7370–7390

Weber H, Chetelat A, Reymond P, Farmer EE (2004) Selective and powerful stress gene expression in *Arabidopsis* in response to malondialdehyde. Plant J 37:877–888

Weigel HJ, Manderscheid R (2012) Crop growth responses to free air CO_2 enrichment and nitrogen fertilization: rotating barley, ryegrass, sugar beet and wheat. Eur J Agron 43:97–107

Wilkinson S, Mills G, Illidge R, Davies WJ (2012) How is ozone pollution reducing our food security? J Exp Bot 63:527–536

Xu Z, Jiang Y, Zhou G (2015) Responses and adaptation of photosynthesis, respiration, antioxidant systems to elevated CO_2 with environmental stress in plants. Front Plant Sci 6:701–717

Yamaguchi M, Watanabe M, Matsuo N, Naba J, Funada R, Fukami M, Matsumura H, Kohno Y, Izuta T (2007a) Effects of nitrogen supply on the sensitivity to O_3 of growth and photosynthesis of Japanese beech (*Fagus crenata*) seedlings. Water Air Soil Pollut 7:131–136

Yamaguchi M, Watanabe M, Iwasaki M, Tabe C, Matsumura H, Kohno Y, Izuta T (2007b) Growth and photosynthetic responses of *Fagus crenata* seedlings to O3 under different nitrogen loads. Trees 21:707–718

Zeng J, Sheng H, Liu Y, Wang Y, Wang Y, Kang H, Fan X, Sha L, Yuan S, Zhou Y (2017) high nitrogen supply induces physiological responsiveness to long photoperiod in barley. Front Plant Sci 8:569–580

Zinta G, AbdElgawad H, Domagalska MA, Vergauwen L, Knapen D, Nijs I et al (2014) Physiological, biochemical and genome-wide transcriptional analysis reveals that elevated CO_2 mitigates the impact of combined heat wave and drought stress in *Arabidopsis thaliana* at multiple organizational levels. Glob Chang Biol 20:3670–3685

Chapter 6
Conclusions and Future Prospects

Contents

1 Conclusion

Ground level O_3 is a widely distributed atmospheric pollutant, likely to threaten the global food production in near future. Bing a secondary pollutant, O_3 is formed from a series of complex photochemical reactions involving the primary pollutants (NOx and VOCs), depending upon the meteorological variables prominent in the region. Background concentrations of O_3 in mid latitudes of northern hemisphere during 19th century have doubled to about 30–35 ppb and have since then increased by another 5 ppb to reach 35–40 ppb. Although O_3 formation is actually a regional phenomenon, the recently developed concept of intercontinental O_3 transport has given a global dimension to O_3 pollution problem. Regardless of the strict air quality legislations implemented to check the emission of O_3 precursors in countries like USA, their background O_3 concentration still show continuously increasing trend. This unexpected trend of background O_3 is actually attributed to the intercontinental transport of O_3 and O_3 precursors formed/emitted from South and South East Asian regions. The existence of intercontinental O_3 transport suggests that the increasing levels of O_3 from one continent may partially negate the efforts taken to control O_3 pollution in another.

A more intensified problem of Northern Hemisphere, i.e. Europe, North America and South East Asia, increasing O_3 pollution has led to severe damage to important agricultural crops. On the global scale, O_3 pollution has the biggest impact in South and South-East Asia in terms of crop production losses. O_3 pollution originating in South- East Asia is responsible for the loss of 6.7 million tonnes of rice per year globally. In a study done by Stockholm Environmental Institute (SEI), it was found that

(i) Asian pollution is responsible for worldwide losses of 50–60% wheat and 90% rice.

© Springer International Publishing AG 2018
S. Tiwari, M. Agrawal, *Tropospheric Ozone and its Impacts on Crop Plants*,
https://doi.org/10.1007/978-3-319-71873-6_6

(ii) North American pollution contributes maximum towards the productivity
     losses of maize (60–70%) and soybean (75–85%).

O_3 enters the plants through stomata and forms highly oxidizing ROS in the apoplastic sub stomatal cavity. The intrinsic antioxidant defense machinery works to
scavenge the ROS generated via O_3 exposure. Plants are able to detoxify ROS generated due to low concentration of O_3 but only to a certain threshold level. Under
prolonged and severe O_3 stress, the ROS generated far exceeds the scavenging
capacity of the plant's defense system. The surplus ROS now targets the membranes
leading to the peroxidation of their lipid components, disturbing the symplastic configuration of the cell. The ROS generated in the chloroplasts target the photosystems
present on the thylakoid membranes causing serious damage to light reaction,
reducing the yield of light reactions. Additionally the capacity of the photosystems
to trap light energy also registers a significant decline under O_3 stress.

Stomatal movement also plays a significant role in determining O_3 injury in
plants. Opening or closing of stomata determines the O_3 flux entering the leaf's
interior. The stomatal closure observed in the plants upon exposure to O_3 may be
either a direct effect of O_3 injury to the membranes of the guard cells, or it may be
an indirect effect wherein the stomata close because of the increasing internal CO_2
concentration which accumulates due to the damaged photosynthetic machinery.
Stomatal closure, however, is an adaptive feature of the plants to avoid O_3 injury by
minimizing O_3 flux in the leaf. Stomatal movement also plays an important role in
determining the sensitivity of different crop cultivars towards O_3 stress. It has been
shown that resistant cultivars show more decrease in stomatal conductance than the
sensitive cultivars under similar O_3 exposure conditions. Several studies have
reported that the modern crop cultivars are more sensitive to O_3 injury compared to
older ones, which is mainly attributed to the increased stomatal conductance (hence
O_3 flux) in the modern crop cultivars.

It has been reported that regions of China, Japan, India, Central Africa, USA and
Indonesia which are the important centres for the production of crops like wheat,
rice and soybean will continue to experience much higher O_3 concentrations in near
future. Studies have shown that global yield reductions due to present day O_3 concentration ranged from 2.2–5.5% for maize, 3–4% for rice, 3.9–15% for wheat and
8.5–14% for soybean. Under the current legislation scenario by 2030, the global
increase in relative yield loss for wheat (4%), soybean (0.5%), maize (0.2%) and
rice (1.7%) is expected with countries of South and South- East Asia being more
severely affected. By 2020, O_3 induced yield reductions in China are predicted to be
18–21% for soybean, 29.3% for spring wheat, 13.4% for winter wheat, 7.2% for
corn and 5% for rice. In India, an increase in the relative yield loss from 2000–
2020 in the range 5.4–7.7% is predicted.

Since the formation of O_3 in the troposphere depends upon multiple factors, air
quality regulations do not prove to be quite effective when mitigation of O_3 induced
injury in plants is concerned. For the proper accomplishment of alleviation of O_3
injury, certain other strategies should be adopted which include (i) use of specific
agronomic practices like nutrient amendments and CO_2 fertilization, (ii) screening

of O_3 resistant cultivars and their promotion for agricultural use in areas prone to high O_3 concentration, and (iii) modification of plant's metabolic processes through genetic technology, so as to improve photosynthetic efficiencies. However, these concepts are still in their preliminary stage and requires further research. Several studies related to cultivar sensitivity with respect to O_3 have clearly indicated a better performance of resistant cultivars. Development of efficient policies promoting the use of resistant cultivars will indeed help in minimizing O_3 injury to some extent. Nutrient amendments and CO_2 fertilization have also given some fruitful results and can be recommended for O_3 mitigation studies. CO_2 fertilization provides extra carbon (biomass) to the plants, which can be allocated towards the development of their reproductive structures which result in improved yield parameters. Nutrient amendments compensates for the biomass which is utilized by the plants in repairing the damage caused by O_3.

Climate change influences the tropospheric O_3 budget by bringing about changes in the meteorological variables and in the rate of emission of O_3 precursors. Variations in temperature, humidity and incident solar radiations and increased emissions of NOx and VOCs are the important features of climate change which affect O_3 formation. In view of the present day climate change scenario, O_3 concentration is likely to show additional increase in near future. Summertime daily (8 h) maximum O_3 is expected to increase at a rate of 0.3 ppbyr^{-1}. In addition, increasing CO_2 levels in the troposphere is an important component and significantly affects the plant performance. Elevated CO_2 not only provides extra carbon, but also controls few biophysical processes such as stomatal movement. Stomatal flux, which determines the quantity of O_3 entering the plants is highly dependent on climatic variables like temperature, humidity and atmospheric CO_2 concentration. As such, the amount of O_3 uptake by plants is indirectly influenced by climate change.

O_3 induced crop loss assessment becomes more significant since O_3 concentration is found to be more in the rural and suburban areas rather than the urban centres. As most of the agricultural land is concentrated in the rural and suburban areas, O_3 impacts on production losses are intensified. Agriculture has played an important role in the development of human civilization and has a major role in determining the socio- economic status of the population, especially in countries where it is the main source of livelihood. With the predicted increase in human population to 9 billion in 2040 from 6.8 billion in 2009 (US Census Bureau, 2009), O_3 pollution will be a key factor in affecting the global food security in the coming times.

2 Future Prospects

The significant role that O_3 is expected to play in deteriorating the global crop production appositely make O_3 a hidden threat to global food security in the coming years. Increasing population load and present climate change scenario further add to the existing problem of O_3 induced production losses. O_3 impact assessment programmes like NCLAN and EOTC have provided a region-wise estimate of O_3

induced crop production and economic losses for USA and European countries, respectively. However, the regions like South and South East Asia, lack such standardized programmes, due to which it becomes extremely difficult to analyze the extent of damage caused by O_3 in these regions. Lack of co- ordinated experimental campaigns and variations in the experimental protocols followed by different research groups fail to provide a common methodology to assess the response of agricultural crops to O_3 pollution. Pilot experimental campaigns developed in these areas over the recent years, co-ordinated by UNEP (United Nations Environment Programme) and APCEN (Air Pollution Crop Effect Network) have given some valuable information regarding O_3 concentrations and its impacts on crop production of the Asian regions. South Asian regions have been predicted to be the major hotspots for O_3 production in near future. Additionally, they contribute a major share in intercontinental transport, and are the important centres for agricultural production of important staple crops like wheat, rice, maize and beans which are sensitive to O_3 stress. These features depict the urgency of initiating crop loss assessment programmes in these regions with immediate effect.

It has been experimentally proved that different cultivars of a species show variable response upon O_3 exposure. If the resistant cultivars are screened and recommended for agricultural practices, negative impacts of O_3 on crop production can be reduced to some extent. Along with the screening of resistant cultivars, there is an urgent need to establish crop breeding programmes to develop resistant cultivars which should have the potential to sustain higher yield even under elevated O_3 stress. Since it has been experimentally proved that the modern crop cultivars are more sensitive to O_3, promoting the crop breeding programmes for increasing crop productivity acquires greater significance. Experimentally proved fact that Asian crop varieties are more sensitive to O_3 stress than the American varieties further intensifies the O_3 pollution problem of this area and calls for an authentic region wise evaluation of the crop production losses.

It has been well established that the production of O_3 and emission of O_3 precursors is no more a problem confined within the country's boundary but has assumed much larger dimensions. The phenomenon of intercontinental O_3 transport has proved the marked influence of South Asian plumes consisting of O_3 and its precursors in determining the concentration of O_3 on the East coast of North America. As the problem of O_3 pollution extends beyond the international limits, there is an imperative need for the implementation of stringent international policies ensuring reduction in the emission of O_3 precursors in developing countries, so that both peak levels and background concentration of O_3 around the globe can be controlled.

The experimental studies done so far, although enriched us with the negative consequences of O_3 on crop production to a large extent. However, there exist a few gaps in our knowledge which restricts the actual estimation of O_3 induced crop production losses. The presence of these gaps or missing data can be due to:

(i) Limited experimentation on different crops and cultivars,
(ii) Limited number of locations, and
(iii) Differential use of experimental protocols.

Owing to these gaps, our knowledge regarding the O_3 impacts on agricultural productivity remain unclear. To fill in these gaps, it is important to establish monitoring networks with a good spatial coverage in rural or remote locations to ensure that the monitoring data provide a clear picture of O_3 concentration to which agricultural crops may be exposed. Further, an assessment of seasonal variation of O_3 concentration is also extremely important. If the growing season of any crop extends over the time when O_3 concentration is higher, crop is anticipated to suffer more production losses. In addition, more field based O_3 experiments should be conducted for newly developing cultivars and their response to O_3 stress should be studied in details.

In addition to the increasing O_3 concentrations, climate change too, is likely to alter crop response. Present day climate change scenario is responsible for changing several atmospheric physical variables such as temperature, humidity, solar radiation etc. The changes in the meteorological variables affect O_3 formation. As such, O_3 concentration is likely to be significantly influenced by the phenomenon of climate change in near future. Climate change not only affects the meteorological variables, it also influences the rate of emission of O_3 precursors. Additionally, climate change has a significant potential to influence the O_3 flux in plants by controlling the stomatal movement. Ozone flux in plants is regulated by opening or closing of stomata which is frequently governed by the meteorological variables. The feature is important from climate change perspective as further increase in temperature in future may promote more O_3 uptake by plants. The aspect of climate change is significant in terms of cultivar specific response of crops as sensitivity/resistivity of cultivars is largely dependent on the stomatal conductance.

3 Recommendations

- International policies which aim at reducing the emissions of O_3 precursors should be framed, so that the negative effects of intercontinental O_3 transport may be minimized.
- The policies aimed at reducing air pollution and air quality abatement measures should be amalgamated, as both combine together to maintain food security.
- Extensive programmes for evaluation of O_3 induced crop loss should be planned, especially in remote areas so that the existing knowledge 'gaps' can be filled.
- More emphasis should be given to the assessment of O_3 impacts on food security in South and South Asian regions which are prone to higher O_3 concentrations in near future and are important centres for cultivation of main staple food crops like wheat, rice, maize and beans.
- Experiments focusing on the screening of O_3 resistant cultivars should be promoted which can be further recommended for reducing O_3 induced crop yield losses.

- Additional field based experiments should be conducted for current and newly developing cultivars, with more emphasis on regionally important staple food crops.
- Crop breeding programmes should be proposed with the purpose of improving the photosynthetic efficiencies of existing crop cultivars and developing new O_3 resistant varieties.
- More experiments are required to be designed to study the interactive effects of climate change and O_3 on the physiology and productivity of crop plants.
- Development of region specific flux models for improving the accuracy of predicting the stomatal response of the plants.
- Agronomic practices like planned soil nutrient amendments should be popularized among the farmers, so that O_3 induced injury in plants can be ameliorated to some extent.
- Various non- governmental organizations (NGOs) working in the field of environmental protection should be motivated to develop a social awareness among the populations which are otherwise ignorant of the serious consequences of ground level O_3 pollution.

Index

© Springer International Publishing AG 2018
S. Tiwari, M. Agrawal, *Tropospheric Ozone and its Impacts on Crop Plants*,
https://doi.org/10.1007/978-3-319-71873-6